Global Disease Eradication

The Race for the Last Child

Global Disease Eradication

The Race for the Last Child

Cynthia A. Needham and
Richard Canning

ASM Press • *Washington, DC*

ASM Press
American Society for Microbiology
1752 N Street, N.W.
Washington, DC 20036-2904

Library of Congress Cataloging-in-Publication Data

Needham, Cynthia, 1946–
Global disease eradication : the race for the last child / by Cynthia A. Needham and Richard Canning.
p. ; cm.
ISBN 1-55581-225-2
1. Epidemics—Prevention. 2. Communicable diseases—Prevention.
[DNLM: 1. Communicable Disease Control. 2. Malaria—prevention & control. 3. Poliomyelitis—prevention & control. 4. Smallpox—prevention & control. WA 110 N374g 2003] I. Canning, Richard, 1960– II. Title.

RA651.N44 2003
616.9'045—dc21

2003004484

10 9 8 7 6 5 4 3 2 1

Address editorial correspondence to: ASM Press, 1752 N St., N.W., Washington, DC 20036-2904, U.S.A.

Send orders to: ASM Press, P.O. Box 605, Herndon, VA 20172, U.S.A.
Phone: 800-546-2416; 703-661-1593
Fax: 703-661-1501
Email: books@asmusa.org
Online: www.asmpress.org

Contents

Acknowledgments

This book turned out to be a very special collaboration for us, one that would not have happened without the influence of a friend and colleague, Rita Colwell, Director of the National Science Foundation, who introduced us. We are grateful for that introduction. We are also grateful to all the friends, family, and professional colleagues who helped us along the way. Their critical comments, valuable insights, and support helped to make this a better book.

We would particularly like to express our appreciation to those scientists who allowed us to interview them and who willingly shared their experiences with us.

Eradication: a Prologue

In the spring of his 13th year, a young boy in Dubuque, Iowa, developed a rare and deadly neurological disorder called subacute sclerosing panencephalitis. As his adoptive parents stood by and watched helplessly, their son became increasingly withdrawn and confused. His grades slipped; he lost much of his language and mathematical skills; he even lost the ability to follow simple directions. Over the weeks he gradually lost control of his muscles, falling to the floor with increasing frequency. In the end, amid the nurses and doctors at one of Iowa's most sophisticated medical centers, he slipped quietly into a coma and died.

The young boy's death could have been easily prevented. His disorder was a complication of measles, a disease he contracted sometime during the first four years of his life when he lived in an orphanage in Thailand. Millions of other children around the world contract measles every year, and at least a million of these children die. Although it is uncommon in the United States, measles is the eighth most common cause of childhood death in the world. Most of these children die in countries where parents are too poor to afford vaccinations and where there is limited or no access to health care. Increasingly, however, American children are at risk, as parents refuse to have their children vaccinated. These parents don't refuse because they are poor. They refuse because they fear complications from the vaccine or because their religion forbids vaccination. More ominously, in recent months the vaccine itself has been in short supply and, therefore, unavailable to some. All these events set the stage for a measles epidemic in our own backyard.

But what if it were possible that no child would ever again die from measles? What if it were unnecessary to vaccinate any child against this disease? Such a time may be closer at hand than you might think. Measles is high on the list of diseases that public health experts believe it may be possible to eradicate—that is, to stop all transmission so that the disease no longer exists anywhere in the world. Smallpox was the first and, so far, the only disease ever eradicated, but at least two more are likely to join it in the archives of human disease.

The history of eradication itself is such, however, that it raises serious questions about whether we can and whether we should attempt to eradicate another disease.

As this book makes clear, the eradication of smallpox was a public health victory of remarkable proportions. Nothing like it had ever been done before, and perhaps nothing quite like it ever will again. The point is not to cast doubt on whether society should strive for such an achievement in the future, but instead to begin to indicate some of the complexities that surround the notion of eradication and the reasons we must choose our battles extremely carefully. To the average person, what goes on in a doctor's office often seems mind bogglingly complex in itself. Disease eradication, because it takes the globe as its clinic, faces monster-sized versions of these complexities. The facilities are worse, the money scarcer, the sun hotter, the cold colder, the workers harder to find and to train, and the results harder to verify. Then, should the attempt, against all odds, succeed, the danger is always present that two or three decades later, someone who thinks he's got the voice of God in his head might undo the work. More than anything else, eradication is a contingent sort of work, contingent especially on geopolitics, cultural values, and luck.

Eradication programs are as delicate as a child's health. This is fitting because, like much of international public health, eradication is built around the child. It isn't going too far to say that without suffering children, there would be no disease eradication. On one hand, children are the main victims of many infectious diseases—malaria, for example, or polio or measles. Adults can suffer and die from these illnesses, too, of course—the world's last victim of smallpox was an adult—but in general they strike early. And because infectious diseases are prominent among the few diseases that can be eradicated, children are, therefore, also the prime beneficiaries of eradication.

Children are vital to eradication in another sense as well: as the chief means by which it is sold to the world. We have even chosen to shape this book around that image in order to dramatize the human consequences of eradicating a disease (or failing to). While *Global Disease Eradication: the Race for the Last Child* focuses on all aspects of disease eradication, the image of the child remains most powerful. The desire to make certain that no child will suffer the consequences of a particular disease ever again—that no parent will suffer the loss of a child because of that disease—makes the race to find that last child, to immunize him or treat her, the most compelling argument for eradication. For that reason, the face of eradication is a child's face to the world at large.

Children are powerful motivators. Consider the March of Dimes, founded to raise money for victims of polio and for research into a cure. Its publicists used the most heart-wrenching images they could find, and these were almost always images of children struggling in braces. The forceful personality of Franklin Roosevelt undergirded the effort, and his experience made stopping polio seem like a matter of national, even international, importance. Still, the pictures on the March of Dimes posters were, more often than not, pictures of children—not presidents or prime ministers, just little kids who would never kick a ball again. The images were extremely effective. Hundreds of millions of dollars were raised and disbursed by the March of Dimes, and in what seems like almost no time at all, an effective vaccine was developed, tested, approved, and put to use.

Today, the number of cases of polio has shrunk almost to nothing; in the next few months, the number will probably reach zero. Although this dramatic reduction could never have been achieved without an effective vaccine, the vaccine alone is not enough. To take that vaccine to the poorest corners of the world, a worldwide campaign was required. And to get that campaign organized, funded, and moving, and to keep it moving, children were crucial. Children are the most effective way the proponents of eradication make their case, particularly to the public at large. To health ministers, the organizers of an eradication campaign might talk about cost/benefit ratios; they might discuss among themselves the technical innovations that make the campaign feasible. But to the volunteer in the field, giving her time and risking her health, to the suspicious parent who has never even heard of vaccination, to the consumer back home, whose donations are an increasingly important source of funds, the talk is more often about saving children.

Concern for children runs so deep, in fact, that civil wars and guerrilla actions have been interrupted so that eradication efforts can proceed. During the late 1960s, for example, Nigerian troops carried smallpox vaccine halfway out onto a bridge and then withdrew so that their Biafran opponents could collect it. Of course, the smallpox eradication campaign did not bring peace to Nigeria or anywhere else. Examples like this, however, sometimes make it seem as though, because it concerns itself so directly with children, eradication operates outside of politics, above the sectarian issues that move one set of people to try to kill another.

This both is and isn't true, as this book shows. Eradication cherishes an ideal—freedom from disease—that makes sense to most people and that transcends mere ethnic or national divisions. Often this message gets across, even in the most difficult of circumstances. On the other hand, eradication, or rather an eradication program, is also a shambling, bureaucratic reality that falls under the control of the highly political World Health Organization. A great deal of money and prestige are at stake, and, therefore, eradication is vulnerable not only to world events—coups, floods, famines, wars against terrorism—but also to politics of a more banal, institutional kind.

The tensions between these idealistic and realistic elements are traced in the sections that follow. Three global campaigns—the races to eradicate malaria, smallpox, and polio—tell the story of disease eradication and herald its future. Each campaign was interesting in its own right, but more importantly, each shaped the international programs that will follow in our race to save the last child.

Malaria and the Magic Bullet

We must learn to shoot microbes with magic bullets.

PAUL EHRLICH

And now alas for mosquitoes! Malaria must be wiped from the earth. Malaria can be destroyed!

DE KRUIF, *MICROBE HUNTERS* (1926)

The important fact is not that there are unsolved, and at present, unsoluble [sic] problems of malaria in the world. It is that there are thousands of communities oppressed for ages by the disease that could be obliterated with a little thought, energy and money, which would not be difficult nowadays to assemble and apply.

HACKETT, *MALARIA IN EUROPE* (1937)

If you want to destroy the mosquito, you must learn to think like a mosquito.

COLBOURNE, *MALARIA IN AFRICA* (1966)

In August 1986, a resident of Carlsbad, California, was diagnosed with malaria. His case alerted local public health authorities in San Diego County and then an epidemiology team at the Centers for Disease Control in Atlanta, Georgia, that something was amiss in this coastal region of southern California. Although a small number of cases of malaria occurs every year in the United States, virtually all the victims become infected in regions of the world where malaria is still a major problem. Malaria was eliminated from the U.S. over four decades ago.

Yet all the evidence indicated that this longtime resident of Carlsbad had become infected locally with the dangerous parasite, setting off alarm bells within the public health community. He had not traveled outside the United States. Investigators discovered female *Anopheles* mosquitoes, the only kind of mosquito known to transmit malaria from one human to another, breeding in the marsh across the street from his house. The same investigators identified an additional 26 people from the area, all of whom had acquired malaria within San Diego County that summer. These 27 individuals represented the largest outbreak of malaria locally transmitted in the United States since malaria was officially declared eliminated from North America in the 1950s.

Malaria was discovered again in suburban New Jersey in 1991, in New York City in 1993, in Houston, Texas, in 1994, and in Loudoun County, Virginia, in 2002. These outbreaks were even more notable because they took place in populous centers rather than in the usual rural settings, like San Diego County, where local transmission had been spotted in the past. The *Anopheles* mosquito rarely inhabits densely populated areas. Further, the epidemics in New York and New Jersey were geographically remote from Mexico, where local transmission of malaria still occurs and where importations into the U.S. are rare but still possible.

The circumstances around outbreaks like these must be extraordinary. For transmission to occur, a female *Anopheles* mosquito must feed on an infected person, survive long enough for the parasite to mature in its gut, and then find an available new host to infect. In each outbreak, these conditions were met. *Anopheles* mosquitoes and a population of immigrants or travelers from countries with malaria came into contact. In each outbreak, the weather was warmer and more humid than usual, conditions favoring the expansion of the mosquito population and the rapid development of the malaria parasites they were carrying. In each outbreak, a susceptible population was available to complete the cycle. Each outbreak served as a reminder that this disease can reappear, even in a region of the world where we have eliminated it, when unusual or changing circumstances provide the opportunity.

Experience is teaching us that there may be many more such opportunities, some of which are products of our own technology. As people travel more and more easily around the globe, the likelihood grows that some of them will bring malaria parasites home in their blood. These unsuspecting travelers, along with visitors or recent immigrants from countries with malaria, serve as reservoirs of the parasites, a source from which new outbreaks can occur. Small numbers of cases caused by local transmission such as the ones in California, New York, New Jersey, Texas, and Virginia continue to appear almost every year in the U.S., in Canada and certain countries in South America, in Europe, and in other places where malaria has long been considered eliminated.

Technology has also expanded the range of the malaria "vector" itself, the parasite-carrying *Anopheles* mosquito. *Anopheles* needs a warm climate with abundant moisture and still bodies of water in order to breed. Historically these environmental requirements restricted *Anopheles* to specific geographic regions, but the mosquito has recently become a regular world traveler, too, stowing away on aircraft, moving literally halfway around the world in a day. Upon arrival, it can transmit the parasite to people living around the airport who have never even left home.

Changing land use through agricultural practices or other development activities creating new mosquito breeding sites also threaten to spread the *Anopheles* into previously uninfested areas. Significantly, some scientists warn that global warming and its attendant climate changes may pose an even greater threat, increasing the area where the mosquito and parasite thrive and perhaps heralding widespread resurgence of malaria in the temperate areas around the globe.

World events often magnify the threat of malaria. Large population movements, like those occurring in sub-Saharan Africa as a result of civil unrest and wars, bring whole new populations into contact with the parasite. People who normally live in malaria-plagued areas develop limited immunity through constant exposure to the parasite; new populations shifting into the area are highly vulnerable to explosive and relatively lethal malaria epidemics.

Malaria, in fact, is still widespread in close to 100 countries. At the beginning of the 21st century, it threatens the health of more than 2.4 billion people every year, and the situation is worsening. Ironically, this is despite important advances in medicine, entomology, and chemistry. We understand more than ever before about how the parasite affects its human host, more about its life inside the mosquito, and more about the mosquito itself. We have both low-tech and high-tech weapons to control it—insecticide-impregnated bed nets and window screens on the one hand, potent insecticides and antimalarial drugs on the other. Scientists are at work on a malaria vaccine, and even on a genetically modified mosquito. Today, however, both the parasite and the vector are thriving.

Even more ironically, malaria continues to be widespread despite a massive global campaign against it that occupied the world from 1955 to 1969. This is the story of that campaign and its aftermath.

Two Children

The year is 1969. The setting is Mali, a small West African country slightly less than twice the size of Texas. A two-year-old child toddles along a muddy road in Bandiagara, a town of 12,000 people in the southern part of this poorest of African nations. The road is lined with houses constructed from wattle and daub, a mixture of mud and straw. Pigs, chickens, and goats wander amid the donkey carts. Cars are rarely seen here. The child laughs and runs to his mother as the rain begins to fall.

It is September, the beginning of the rainy season. With the rains come the dreaded fevers that sicken the people in the community and take the lives of their children. And with the rains come massive increases in the mosquito-breeding pools of water. In Bandiagara, like much of the developing world, climate and economics have conspired to make things easy for mosquitoes. Their bite is a normal feature of life.

Each night, as the two-year-old slips into bed with his mother, the night-feeding mosquitoes descend to take their blood meal, biting mother and son repeatedly as they sleep. Many of these mosquitoes bring with them the deadly *Plasmodium* parasites that cause malaria. The two-year-old is soon the victim of this deadly combination of mosquito and parasite. He becomes infected, like virtually all the people in the community, old and young.

The disease is initially silent as the parasites multiply inside the child's liver cells. Within a week, the malaria parasites are circulating in his blood, and suddenly they begin to take their malignant toll. The child, fretful and fevered, interrupts the quiet of the night with the sounds of his coughing. The warning signs worsen throughout the night, and then in the early morning the seizures begin. His mother is no stranger to this disease, which she knows as *wabu,* but like others in her community, she doesn't associate it with the fevers of the rainy season. She listens apprehensively for the call of a bird, for she believes that if her son cries out at the same time, the bird will steal his spirit and take her son from her.

She knows she must now act quickly. As her son slips into unconsciousness, she carries him to the local healer. Inside her son's brain, red blood cells filled with malaria parasites have gathered, restricting the flow of oxygen. He is a victim of cerebral malaria, a disease that kills over 50% of its sufferers, most of whom are children. The herbal brew used by the traditional healer can't save him. One day he is a normal, happy child, and the next he is gone.

A second child, on the surface, is like the first: the same community, the same mosquitoes, and the same parasite, but a seemingly different disease. Although he, too, is very ill, he's among the fortunate. This child also develops the fever and cough, but he does not make the sudden descent into the unconscious state of cerebral malaria. Instead of dying, he recovers to begin a cycle that will continue throughout his life, as it does for all the surviving villagers.

This child is bitten and infected and treated with every rainy season during his childhood, each bout of the disease a little less severe than the one before, and each bout producing a little more resistance in him for the next attack. By the time he is a young man, malaria will be mostly an inconvenience rather than a specter of death; he will continue to be infected with each rainy season, but his symptoms will be relatively mild. Ironically, in part it is the infection that is the key to his health, at least as far as malaria is concerned. As he's bitten again and again, an almost continuous supply of parasites is maintained in his blood, helping him remain more or less immune to their effects for the rest of his life.

These children are fictitious, but they are not unlike many real children who lived in malarious regions of the world in 1969, at the end of the malaria campaign. They are not unlike many children living in the same regions today. More than 300 million people still suffer from malaria throughout the world each year, over half of them children. Even more disheartening, more than one million children die from malaria each year—about one every 30 seconds.

Like many of the worst infectious diseases, malaria is hardest on those least able, economically or politically, to defend themselves. The poor are more likely to live near the swamps and pools where mosquitoes breed; they are more likely to live in houses that allow mosquitoes to come and go. They have less money to spend on simple preventative measures—bed nets or window screens—and less access to adequate health care. And not only are the poor more likely to suffer from malaria, they are also, because of malaria, likely to remain poor. The debilitating disease diverts manpower from the planting and harvesting upon which subsistence communities depend. It compromises the normal physical and mental development of the children who are the lifeblood of the communities, and it discourages multinational corporations and private investors from establishing projects that might infuse much-needed capital into struggling economies.

The high costs of malaria and the belief that we possessed the technology and will to defeat the parasite once and for all led the public health community in the 1950s and 1960s to attempt the first global effort to *eradicate* a disease. Eradicating a disease—that is, stopping all transmission, pulling it out by its roots—is quite different from controlling or eliminating it from a part of the world. Control and elimination are important steps along the way, but they are not the same; they are not the ultimate public health goal—zero transmission—that defines eradication.

In other words, when the malaria eradication campaign was being considered, it was viewed as a massive effort to lift, once and for all, the burden that the disease places on all nations.

It must have seemed reasonable to many people at the time that "a little thought, energy and money," as one writer put it, was all that was necessary to eradicate malaria. Many of the world's leaders in public health believed fervently that eradication could work, and they devoted huge financial and human resources toward that end for over 14 years. But the men and women who participated in that massive global campaign from the mid 1950s through the 1960s found out otherwise. Although great strides were made in many countries, the disease was not eradicated.

As the recent malaria outbreaks in the U.S. prove, eradication is an all-or-nothing proposition. If a disease has not been eradicated everywhere, it has not been eradicated anywhere—controlled, yes, even eliminated, but not eradicated. As long as the malaria parasite circulates in Mali, Mexico, or anywhere else, the chance always exists that it will show up again in San Diego or New York—or any other place where malaria transmission is no longer thought to occur.

So the story of this first global campaign to eradicate a disease is ultimately a story of failure because malaria has clearly not been "wiped from the earth." It offers a hard lesson in reality—that no matter how firm the commitment of the men and women working around the globe, and no matter how much political will is mustered or how many resources are applied, eradicating most infectious diseases simply is not economically or socially feasible and may not be technically possible. The story of the Malaria Eradication Programme, or MEP as the World Health Organization called it, is worth remembering in its own right, however, perhaps even celebrating, as one of the first times that people from around the globe worked together to achieve a common good. And as is often the case with failure, important lessons can be learned from it.

Swamps, Farms, and Bad Air

Malaria is an ancient enemy. The parasite that causes the disease might even have affected our prehuman ancestors. Recent evidence places the spread of the most severe form of malaria to a time 12,000 to 7,000 years ago in western Africa, when climate changes were triggering changes in human lifestyle. Malaria is identifiable in Chinese texts that date back to 2700 B.C., and by the fifth century B.C.—2,500 years ago—people had begun to understand something about its causes and about ways to fight it.

Hippocrates, the founder of Greek medicine, noticed the connection between stagnant water and outbreaks of fever, and he blamed the fevers on the odor such water produces. The Romans elaborated this idea, gave the disease the name it carries in English today ("*mal' aria*," or "bad air"), and drained marshes and swamps to fight it. This no doubt improved the quality of the air, but the air itself had nothing to do with malaria.

It's all about mosquitoes, and the same water that produced bad air also gave mosquitoes a fertile place to breed. Draining the swamps and marshes fortuitously lowered the number of disease-laden insects and thus the likelihood of acquiring the disease. The role of the mosquito in spreading malaria remained undiscovered until the end of the 19th century, and the importance of surface water management in controlling the disease remains unchanged even today.

A second measure to thwart malaria was also discovered well before we learned malaria's cause and the manner by which it moves from person to person. By 1650, the antimalarial properties of quinine, a drug extracted from the bark of the cinchona tree, were well known in Europe and throughout South America. Drinking the extract made from cinchona bark killed the *Plasmodium* parasites in the sufferer's blood. The drug was not then widely available, but its use provided early evidence of the importance of pharmaceutical strategies for tackling the disease.

Malaria continued to occur unabated throughout most of the world despite these measures. The disease could be found in both temperate and tropical climates up through the middle of the 19th century, reaching as far north as Canada, Norway, Sweden, and Denmark. Then the numbers of people infected in more temperate climates began to decline.

The reasons for the early decline in malaria were mostly happy accidents, the unintended but beneficial effects of long-range changes in farming and demographics. People began to drain swamps—not necessarily to combat malaria, or even bad air, but to reclaim valuable land. The result, of course, was the elimination of the mosquito breeding grounds. Farmers began to use root crops, such as turnips and mangelwurzels, for winter fodder, which in turn made it possible for the farmers to keep larger numbers of domesticated animals throughout the year. As it turns out, many species of *Anopheles* mosquito prefer to feed on animals other than humans. The increased availability of mosquito fodder helped to decrease the probability that a human would be bitten.

Other changes in the way people lived pushed the malaria rate down even further. Machinery began to replace manual labor, and more and more people moved to the cities, further reducing the number of human bodies available for hungry mosquitoes in rural areas. People remaining in the countryside learned to build houses that were more or less mosquito proof. Finally, medical care improved; quinine became more readily available, and its cost fell. More infected individuals received the drug, reducing the life span of the parasite and thus the probability that it would be passed to another mosquito, and thus to another human.

As malaria rates dropped, a few people at the beginning of the 20th century began to believe that the disease could be brought under control.

Escalating Pressure

Efforts over the first four decades of the 20th century specifically aimed at reducing malaria and other mosquito-borne diseases were increasingly successful, adding to the growing belief that malaria could be beaten. The first major battle, commanded by the American General William Gorgas, was waged in the Panama Canal Zone as a part of the massive U.S. effort to build the canal. Earlier French efforts to build the waterway had failed largely because malaria and yellow fever, another mosquito-borne disease, killed so many workers. A French resident of Panama at the time, in fact, had warned, "If you try to build this canal, there will not be trees enough on the isthmus to make crosses for the graves of your laborers."

The U.S. goal was to create a mosquito-free zone around the canal construction. Between 1904 and 1910, American military engineers and "sanitary police" aggressively worked to drain swamps, vastly reducing mosquito-breeding sites. Drainage was coupled with the use of larvicides for the mosquitoes and quinine prophylaxis for the residents. The campaign was successful; malaria and yellow fever were brought under control, and the Panama Canal became a reality.

During the 1930s, the Italian dictator Benito Mussolini engineered another massive effort to eliminate malaria from a part of Italy known as the Pontina. The wetlands of the Pontina begin below the hills of Rome and continue west to the sea, flanking the Appian Way, Imperial Rome's major road to the Mediterranean. The area was heavily populated by, among other things, the *Anopheles maculipennis* mosquito and was one of the most malarious places in the world, so much so that the area was fundamentally uninhabitable. Mussolini, heeding the advice of the malariologists of the time, oversaw the construction of an extensive series of drainage canals. As the Pontina dried, the mosquito population fell, and so did the incidence of malaria. As malaria receded, resettlement became possible, and the towns and farms that had not been populated for more than 2,000 years came back to life.

Although greatly reduced, malaria transmission in this region continued until the late 1940s, when *A. maculipennis* finally disappeared from the Pontina. Ironically, as the human population increased, the amount of water pollution did as well. As this particular species of mosquito breeds only in pristine waters, ultimately *A. maculipennis* died out, and malaria was finally eliminated from the region.

Another important campaign set the stage for the growing eradication movement. *A. gambiae* was discovered in Natal, Brazil, in 1930. Although many species of *Anopheles* can transmit malaria, *A. gambiae* is among the most vicious of all malaria-bearing mosquitoes. It prefers to breed in muddy pools of water; its larvae can even be found in water-filled footprints. It prefers humans to all other animals, and it is particularly long lived, giving it the opportunity to spread the parasite to large numbers of susceptible people. *A. gambiae* is an African mosquito, and it had no business being in Brazil. Its importation, most likely a consequence of the mosquito stowing away in the hold of a French destroyer bringing mail from Africa, was followed by a series of small outbreaks of malaria.

Its discovery wasn't sufficient to move the Brazilian government to action, but later events were. Within a few years, the mosquito had spread into an area along the coast and inland covering close to 18,000 square miles. In 1938, Brazil experienced the worst outbreak of malaria in the history of the Americas. The mosquito infected over 100,000 people, killing nearly 20,000. The government called for a major attack against the vicious vector, one that involved the extensive use of larvicides such as Paris green and the fumigation of houses, cars, trains, and any other places where the mosquito might hide. The man who orchestrated this initiative was Fred Soper, and his determination and organization led to the successful elimination of the *A. gambiae* vector. With the elimination of the mosquito vector, malaria came under control in Brazil.

Soper was to prove most influential in the ultimate decision to commit to a global eradication campaign as well as in its fundamental design. One of his colleagues, Don Henderson, who later led his own eradication effort against a different lethal pathogen, describes him as "a giant of a man in terms of recognition, and he was a very determined person with very decided views. It was his belief that if you simply put in really solid management, you could overcome almost anything. He had demonstrated this in fact in Brazil. He actually went in and was successful in eradicating *A. gambiae*. Admittedly, [the mosquito] had certainly not gained a foothold over a wide area, but it had gained a pretty significant foothold. This was really a brilliant achievement."

Such victories suggested to some malariologists that a massive effort against the disease vector, the mosquito, could bring an end to malaria. Others remained skeptical. Some favored attacking the parasite directly, using antimalarial drugs to treat infections so that nothing was left for the mosquito to pick up and transmit. Many questioned the very concept of eradication and doubted that mosquito populations could ever be reduced sufficiently to break the cycle of transmission.

Developments during World War II, however, helped fuel enthusiasm for targeting the mosquito and supplied further evidence that the malaria cycle could be broken. Malaria was taking a heavy toll on the Allied troops fighting in Asia, the Pacific, and northern Africa. American and Allied armies were losing as many men to malaria as they did to combat. At one point in the war, an entire marine division of 17,000 had to be withdrawn from the Pacific theater because over two-thirds of the troops were suffering from the disease. According to accounts from Bataan, close to 90% of the men holding the island were infected with the parasite. General Douglas MacArthur was quoted at the time, saying, "This will be a long war, if for every division I have facing the enemy, I must count on a second division in the hospital with malaria, and a third division convalescing from this debilitating disease."

Finding a way to control the disease assumed top priority among the military. The biology of malaria offered those developing the military's antimalaria strategy with two potential targets—the mosquito and the parasite itself—each of which had been successfully targeted in earlier campaigns.

Two major developments—a new antimalarial drug and a new insecticide—offered potent weapons against the parasite and the mosquito, respectively. Both helped reduce the debilitating influence of malaria on the Allied war effort, although the former turned out to be much less significant than the latter.

Quinine had been used to treat malaria since the mid-1600s, but the drug was not particularly good at preventing the disease, and it was difficult to obtain. Quinine is short acting; as a preventative measure it must be taken frequently, but frequent dosing increases its toxicity. In addition, the Allies soon discovered that the supply of quinine was unreliable. The Japanese controlled most of the world's reserve, and they were on the other side. Atabrine—the only alternative at that time to quinine—was highly toxic. The challenge, then, was to develop an antimalarial drug that was plentiful, effective, and safe, and that could actually prevent the disease.

A new antimalarial drug called chloroquine ultimately offered a successful strategy against the parasite. Originally developed by a German pharmaceutical company, the drug came into U.S. hands after the fall of Tunis in 1943, compliments of the Vichy physicians who had run clinical trials with it earlier in the war. With further development, chloroquine provided a new weapon for preventing malaria. Unlike quinine, it was relatively inexpensive, was long acting, and had limited side effects.

The most fearsome new weapon, however, was dichloro-diphenyl-trichloroethane, a compound that the world came to know as DDT. Developed in Switzerland by Paul Müller during the late 1930s, DDT was thought to be the ultimate insecticide, the consummate bug killer, and it promised—at least according to some malariologists—to change the world. Müller later won the Nobel Prize for his discovery.

DDT quickly became the weapon of choice against the mosquito. The synthetic insecticide was highly effective, inexpensive, and as far as anyone knew at the time, safe. The other property that made DDT unique was its staying power. No other insecticide has the same longevity. And the insecticide swiftly proved its value. Malaria cases declined dramatically when DDT "bombs" began to be dropped onto malarious areas where Allied troops were deployed. Some might argue that the insecticide had as much to do with victory in the Pacific as any other single factor in the war.

Learning from the Past: Hard Lessons, Hard Work

The famous malariologist Paul Russell had early remarked, "What a paradox! Man, with his incredible machines and his streamlined science, stricken each year in millions, because he fails to outwit a mosquito carrying Death in its spittle!"

By the mid-1940s, as an increasingly serious movement to eradicate malaria was taking shape, it finally seemed that both technology and political will had come together to break that paradox. On the technological side, DDT had proved itself as a powerful weapon against the vector *Anopheles* mosquitoes. Chloroquine was so effective in deterring the parasite that it had become the equivalent of table salt for colonials living in malarious countries. The drug had been used to treat thousands of victims with the disease. On the political side, several powerful players had emerged. The U.S. government created the Communicable Disease Center (now the Centers for Disease Control and Prevention) in 1946 primarily for the purpose of fighting malaria. More significantly, international health agencies with the scope and expertise to coordinate a global campaign had been established—the United Nations International Children's Emergency Fund (UNICEF) in 1946, followed by the World Health Organization (WHO) in 1948.

Added to these developments was a theoretical conviction that victory over malaria was achievable. The balance of factors that make malaria transmission possible is extremely delicate. George MacDonald, of the Ross Institute in London, had constructed a mathematical model demonstrating that a serious reduction in the number of vector mosquitoes would make it difficult for the malaria parasite to reproduce itself. To eradicate *malaria,* in other words, it was not necessary to eradicate *mosquitoes;* it was only necessary to kill a lot of them and thus upset the balance. To reduce the number of vectors is to reduce the number of bites, which reduces the number of parasites in the blood of a human host; this in turn reduces the number of parasites a mosquito can ingest from that host, and so on. Further, if vector mosquitoes lived shorter lives, then the parasite, developing in the gut of the mosquito, would have less time to mature.

The proponents of eradication, led by Fred Soper, focused increasingly on the mosquito, and for Soper, DDT was a godsend. After the end of the war, he devised a test for the insecticide to see whether he could replicate his earlier successes in Brazil. His test ground was Sardinia, a highly malarious island off the coast of Italy. In what had become classic Soper style, he organized the Sardinia attack according to a rigorous protocol, dividing the rocky and mountainous island into geographical units, each one encompassing just enough terrain that a single man carrying a spray canister full of DDT could cover it in a week. He hired 33,000 people, acquired 287 tons of DDT, and set off with military precision to attack *Anopheles labranchiae,* the resident vector on the island. Over the course of four years, Soper's sprayers applied DDT to 337,000 buildings. When he began his campaign in 1947, there were more than 75,000 cases of malaria. When he finished in 1951, there were nine.

In retrospect, at least one or two other strategic components shared responsibility for the success of Soper's Sardinia campaign. *A. labranchiae* is a mosquito that breeds both in open water and in the weeds surrounding streams and marshlands. In addition to spraying tons of DDT, workers also cut back thick vegetation and drained huge tracts of marshland. Complementing these efforts was the fact that Sardinia is an island, relatively small and geographically isolated. No one bothered to find out which of these factors contributed the most to the dramatic drop in the mosquito population and thus in the malaria rates. At the time, most were happy to give the credit to DDT.

Almost simultaneously, another campaign was taking place in the U.S. Called the National Eradication Program, the campaign was based on a strategy of indoor residual spraying with DDT. Five years after the program's beginning in 1947, malaria had virtually disappeared from the U.S. Better surface water management and improved housing were pre-existing and contributing factors, and somewhat later it appeared that transmission had already stopped by the time spraying began. Again, however, the success was attributed to DDT.

Soper became convinced that he could drive malaria from the face of the earth by targeting only the mosquito with DDT and putting in place the same administrative and supervisory methods he had used in previous campaigns. He began to make that argument with increasing fervor. His argument in favor of DDT was a compelling one. The insecticide was not only cheap and effective, it was fast, and speed was important for at least two reasons. One was political. Even though DDT itself was cheap, the eradication program was likely to cost millions of dollars. Because eradication would cost so much, there was a need to convince governments and funding agencies that the end of malaria was more or less in sight. They could be promised immediate and permanent benefits in return for heavy, if short-term, investments. The other reason that speed was important was biological: from malaria control programs in Greece, scientists knew that mosquitoes developed resistance to DDT within about five years of the time it was first used, and resistance to DDT was spreading. If a massive campaign was not completed quickly, DDT would lose its effectiveness, and an important opportunity would be lost forever.

These arguments laid aside the other lines of attack that could have been used, each of which had proved valuable within individual efforts. Against the mosquito, draining swamps and other breeding areas and building mosquito-proof houses had also been shown effective. These solutions were viewed as too expensive in developing countries where malaria predominated, as well as too slow. Chloroquine, the drug that prevented malaria, was never really considered seriously for use in a major campaign conducted primarily in developing countries. The drug was not that cheap, and there was justifiable skepticism about the ability of the health care infrastructures in many of the affected countries to deliver the drug appropriately. In the opinion of many, compliance would present insurmountable challenges.

Optimism in some quarters for a malaria eradication campaign ran high, and momentum grew. This was the age, after all, when medical technology was on the ascendancy, and infectious diseases seemed all but conquered. New drugs such as penicillin and vaccines such as those against smallpox and polio convinced observers that miracles could happen. Malaria fit squarely into that thinking. One expert on tropical medicine recalls the advice he received in the early 1950s from one of malariology's giants, H. E. Short: "Desowitz, malaria is about to be totally eradicated, and you will never make a career, let alone a living, from it."

Ultimately, political events sealed the decision of the world's public health community to go forward with a global campaign against malaria. In 1953, a Brazilian malariologist named Marcolino Candau became director-general of the WHO. Candau had worked with Fred Soper in Brazil during the drive to eradicate *A. gambiae* and was influenced by his mentor's arguments. With Candau's help, Soper convinced the members of the World Health Assembly (WHA), the policy-making arm of the WHO, to commit to a global campaign. In May of 1955, the WHA finally voted that "The WHO should take the initiative in a program having as its ultimate objective the worldwide eradication of malaria." The initiative was to be known as the Malaria Eradication Programme, or the MEP.

The MEP's strategy depended on a plan elegant in its simplicity—a plan that depended almost exclusively upon a narrow, vector-centered approach. The mosquito was the culprit—the transmitter of the disease—and the mosquito was vulnerable to DDT. Eradication was merely a matter of bringing these two components together. The basic idea was easy to sketch: use enough DDT to kill enough vector mosquitoes to halt the transmission of the parasite for three to five years, during which time all active cases of malaria would burn themselves out. The human source of the parasite would disappear, and the disease would be defeated once and for all—in a word, eradicated.

The plan had three phases. The first year was one of *preparation,* when the structures to be sprayed with DDT were to be mapped in detail, supplies ordered, and personnel recruited and trained. The second phase, the *attack,* was to begin in 1956 and last three to five years, with the teams spraying all the interior walls of the structures in their region with DDT every six months. In the latter part of the attack phase, surveillance agents were to visit every house once a month to look for suspect cases, taking blood samples from anyone who had symptoms of the disease. The suspected cases were to be treated and additional spraying conducted. By 1961, during the third phase, consolidation, routine spraying was to be discontinued. Then and only then was intensive surveillance to be used to find and treat the few remaining cases of disease, and spraying conducted to eliminate the remaining mosquitoes from the immediate area. The MEP was well funded and well organized; the eradicators expected to fight malaria for the requisite number of years and were prepared to do so.

Fourteen years later, with malaria still a major problem throughout the tropical world, the MEP was called off. In 1969, the WHA shifted its strategy from eradication to control, concluding that, in spite of technology and mathematical certainties, in spite of many hundreds of millions of dollars of funding, eradication of malaria in much of the world was simply not possible.

The global state of affairs at that time tells the story. In sub-Saharan Africa, the malaria situation was virtually unchanged. After fourteen years, a serious effort hadn't even gotten under way. The logistics of organizing a tightly managed campaign in a part of the world where a large percentage of the population lived in remote villages were overwhelming. These challenges coupled with the presence of the vicious *A. gambiae* vector, which bred year-round, forced all but the most ardent to agree early on that virtually nothing could be done against malaria for most of the African continent. The most dispiriting places, however, were the countries where significant efforts and gains had been made—India, for example, and Sri Lanka—for even there malaria was returning with a vengeance. Finally, mosquitoes resistant to DDT had appeared among the mosquito populations where spraying had been heaviest, effectively disarming the eradicators sole weapon of attack.

The retreat of the MEP cast a long shadow over the very notion of eradication. Confidence in the technical and administrative abilities of the WHO to conduct such a campaign was seriously eroded. Many governments and international funding agencies had invested heavily in the MEP. In light of the results—resurgent malaria and insecticide-resistant mosquitoes—it was hard to argue that the investment had been entirely worth it. Such attitudes compromised the planning and execution of both ongoing and future WHO projects. Perhaps most significantly, the failure of the MEP illustrated the daunting complexity of disease eradication, leaving behind lessons that still influence thinking three decades later.

The history of the MEP reveals three crucial factors that must be considered when any eradication initiative is being contemplated. First, all involved must clearly and completely understand the biology and epidemiology of the disease. Second, all must clearly and completely understand the social aspects of the undertaking, from the perspective both of those carrying out the eradication effort and of those directly affected by it. Third, the strategy and the organization must be flexible enough to identify and respond to variations in and discoveries about the biologic and social aspects as the program unfolds. Examination of each of these critical factors makes clear how much more is involved in eradication than the mere killing of viruses, bacteria, parasites—or mosquitoes.

Biology Plays Its Hand

The biology of malaria presented complexities that were little understood at the time. Because DDT was believed to kill mosquitoes the same way everywhere, it was easy to conclude that eradication itself would proceed the same way everywhere. Wall spraying thus became almost the exclusive approach to vector control in all participating countries. And DDT did work. Not only did it kill mosquitoes, its potency was quite long lasting; a surface sprayed with DDT didn't have to be resprayed for months. This approach, however, failed to take into account the complex behavior of the mosquitoes involved and the variety of interior "walls" that it would be necessary to spray.

Spraying interior walls targets exclusively those mosquitoes that prefer to feed indoors and then rest for a time on the nearest vertical surface. While this approach works well where the vector mosquito wall-rests, this behavior is by no means the sole pattern for all of the 50 or so species of *Anopheles* known to transmit the disease. For example, the primary malaria vector in the Southeast Asian jungles is *A. dirus.* The mosquito feeds in the early evening and then rests outdoors. Residual wall spraying has virtually no effect on this mosquito population.

Further, the number of different malaria-transmitting species within a region influences the success or failure of residual wall spraying with DDT. The most common species in an area, called the primary vector, is easily recognized and relatively easily killed. Secondary vectors, however, are often hard to identify even when you're looking for them. They occur in smaller numbers, and their feeding patterns and other habits may well be different, so an attack successful against the primary vector may not affect secondary vectors as much. Their presence introduces an often unanticipated player in local transmission of the disease. Where a secondary vector with different behavior patterns remains after the primary vector is reduced in number, malaria rates often fall but then remain stable.

Residual DDT didn't always achieve the goal even against wall-resting mosquitoes. In the savannas of sub-Saharan Africa, for example, the intensity of transmission was so high that reducing the mosquito population by as much as 99% failed to make a dent in the number of people infected. The principal vectors in this region, *A. gambiae, A. funestus,* and *A. arabiensis,* are among the most efficient transmitters of the disease in the world. They are present in huge numbers, particularly during the rainy season. In many regions virtually everyone is infected at least part of the year, so almost all the female mosquitoes will take a blood meal during their life span from someone who has malaria. These mosquitoes have especially long lives as well, giving them an extended opportunity to transmit the parasite. Wall spraying was not sufficient to break the cycle of infection in these circumstances.

After vector control projects undertaken in the 1960s in Nigeria, Cameroon, and elsewhere in Africa had failed, the WHO wanted to know whether a more intensive effort might succeed. They chose the 164 villages in the Garki district of Nigeria to make the test. Working together, the WHO and the Nigerian government spared no expense, pouring both manpower and money into the project—over $US 6 million in seven years. They sprayed every hut with insecticide at least every ten weeks. They provided antimalarial drugs to a third of the villagers and treated those with active cases of the disease.

This offensive had a spectacular effect on the mosquito population in the region, and it reduced the number of bites each person received by 90%. Yet the incidence of malaria in the Garki villages remained essentially unchanged. Health workers had not stopped the transmission of malaria; worse, they hadn't even reduced it significantly. The effort was defeated, in other words, not by a lack of resources or will, but by the biology of the parasite and its vector.

The effort in Garki proved, disappointingly but convincingly, that until different tools become available, malaria will remain in sub-Saharan Africa. It proved a larger point as well: that the transmission of malaria, and thus the control of it, depends heavily on environmental, climatic, and biologic factors—all of which must be taken into account when designing an eradication strategy.

Within other regions, resistance to the insecticide became a prominent factor. In some cases, the mosquitoes simply changed their behavior. DDT is an irritant and the mosquitoes just didn't stay on a wall covered with the insecticide long enough for it to have its deadly effect. In other cases, true resistance occurred. Small numbers of DDT-resistant mosquitoes in a very large population are of little or no consequence. When massive numbers of susceptible mosquitoes were killed after wall spraying began, however, the small number of resistant insects all survived and continued to breed, ultimately replacing the original population with vectors upon which DDT had no effect. Even if malaria had been brought under control initially, as soon as the resistant mosquitoes reached sufficiently high numbers, malaria returned to its precampaign levels. This was the case particularly in regions where DDT had widespread agricultural applications.

Finally, not all interior walls are created equal. Where village huts are constructed with mud and straw, for example, the walls tend to melt away during the rainy season. Hence they require frequent repair. The timing of their demise matches the time when mosquitoes are gearing up their breeding capacities; thus, walls sprayed earlier might no longer exist during the season of highest transmission. DDT also deteriorated much more quickly on certain substances, such as the clay used to make adobe walls in Mexico, reducing the level of effective insecticide well before the next visit from the sprayers.

Although some of these factors were known and others should have been predictable, little or no field research accompanied the MEP campaign. Consequently, when biology began to play its hand, the failures that resulted were all too often detected only when the number of malaria cases began to increase.

Social Issues Loom Large

With hindsight, it's clear that though eradication rests firmly on both chemistry and entomology, it depends even more heavily on human beings. Perhaps the chief lesson from the MEP concerns exactly this point: eradication is not something an international health community can do *to* or *for* people. It can only be done *with* them. Compared with understanding and managing the human side of the program, the development of a magic mosquito bullet seems positively easy. During every phase, in fact, eradication of malaria was fatally vulnerable to the carelessness, misapprehensions, and simple resistance of people.

During the attack phase, for example, it was human beings who went laboriously, house by house and hut by hut, spraying DDT. For the MEP, concerned with killing as many mosquitoes as possible, the difference between doing this job well and doing it poorly was enormous. It was during the wall-resting period, according to plan, that the vector would encounter the DDT. This meant that almost every dwelling where a mosquito might feed on a human had to be sprayed, and sprayed properly: a 5% solution of DDT in water; 40 pounds of pressure per square inch; 18 inches between spray nozzle and surface; vertical spraying sweeps, 20 inches per second. Such precision was required in order to leave enough DDT on each square meter of surface sprayed, and it's easy to imagine that even the most responsible spray team might occasionally make a few important mistakes. An irresponsible team could really botch things and so undermine the work in an entire region. In a campaign so vast, where large numbers of workers were needed, and needed fairly quickly, it wasn't always possible to tell the careful from the careless. In a real sense, then, the eradication program rested not so much on the killing power of DDT as on the all-too-human notion of the work ethic.

Similarly, it was humans who allowed these spray teams into their homes twice per year, or refused. As long as malaria workers and local communities cooperated, it was possible to accomplish the primary task of the attack phase—bringing together the mosquito and the DDT. Yet without a fair degree of mutual understanding, such cooperation could be only short-lived. The MEP spent too little time trying to foster that understanding. As one medical historian has noted, because the MEP treated malaria as a single problem with a single solution, it proceeded as if the people who had the disease did not exist.

Many people did not even understand why their homes were being sprayed. Around the world, people have developed a variety of explanations for why malaria occurs, explanations that, as was the case in ancient Rome, may have nothing to do with mosquitoes or the parasite. Among the hill tribes along the Thai-Burmese (Myanmar) border, for example, the disease was believed to come from drinking bad water. In Malawi, villagers not only associated malarial fevers with mosquitoes, they also believed that malaria was caused by getting wet in the rain or getting cold, hard work, spirits or witchcraft, and dirty water and food. Against these beliefs, the idea that an invisible spray on the walls could stop fever may have seemed fairly ludicrous.

People could see with their own eyes, however, that the spray got rid of pests in general, and many welcomed the spraying for that reason alone. As the years passed and the pests began to return, having developed resistance to DDT, it became more and more difficult to see the point of the inconvenience. Nevertheless, as long as the spraying caused no serious problems it generally continued unimpeded.

When DDT began to cause a variety of problems, the MEP found itself in thorny situations, even if people did understand what was at stake. The famous malariologist and raconteur Robert Desowitz detailed the story of villagers in Malaya. The villagers used palm fronds to construct the roofs for their houses. These fronds are the preferred food of a particular kind of caterpillar, which is in its turn the food of a particular kind of wasp. DDT is for the most part an indiscriminate killer, and the most immediate benefit of spraying is the disappearance not only of mosquitoes, but also of flies, spiders, and nuisance insects in general. However, not all insects are affected equally. DDT upset the balance between the Malayan wasps and caterpillars, killing the former but not the latter. This left the caterpillars to proliferate and devour the roofs, which promptly began to collapse. Though the villagers understood the association between controlling mosquitoes and eradicating malaria, this knowledge didn't put the roofs back on their homes. They soon ejected the malaria workers.

In addition, the planners of the MEP failed to understand the complex attitudes that people who live with the disease can have toward it. Many people, for example, didn't fear malaria as much as the WHO assumed they did—or told the rest of the world they did. This is not to trivialize malaria, but to stress again its variety around the globe. In endemic regions, places where almost everyone suffers from malaria, the disease can be relatively mild. Often only a drop of blood viewed under a microscope can distinguish a person with the parasites from one without them; there may be no symptoms at all. This is a mystery that science continues to ponder: why malaria is deadly for some but mainly inconvenient for others. For the MEP, however, it was a mystery with important social consequences, and one that the leadership didn't take fully into account.

A villager who has lived with malaria all of her life and knows that her parents lived with it, and her grandparents, and so on, further back than anyone can remember—such a person may not view malaria as the greatest challenge in her life. Other diseases, such as infectious diarrhea and tuberculosis, have a more prominent impact. No doubt it was difficult to convince her that malaria was public health enemy number one—more difficult than convincing the international funding organizations. Yet it was on her cooperation during both the attack and consolidation phases of the program that eradication depended.

It's important to stress that this qualified response to eradication—neither hostile nor enthusiastic—on the part of the people affected by the disease had nothing to do with laziness, stupidity, childishness, or any of the defects typically attributed to the developing world by the exasperated West. It was a response grounded in experience—the experience of actually living with malaria—and in long-standing custom. Add to these the basic human tendency to view anything new with at least a little suspicion, and the qualified response begins to seem fairly recognizable. How many Americans or Europeans, for example, would allow their homes to be sprayed twice a year with a mysterious poison purported to stop heart disease? How long would they allow it? And how would these numbers change if the organizers and the chemical came from, say, Borneo?

For a less far-fetched analogy, consider the flu. Like malaria, influenza has a distinguished history as a killer. In the pandemic of 1918–1919, the Spanish flu killed 20 million people, twice the number that died in World War I. In 1957, influenza killed 70,000 people in the U.S., and in 1968 it killed 34,000. Each year, an average of 20,000 people in the U.S., most of them elderly, die from the flu, and many more suffer its consequences. Yet how many Americans shake in their shoes at the thought of the flu? How many take even modest precautions against it? Fewer than one-third of Americans receive a flu shot each year, even though these shots are often provided free of charge.

It would be easy to put this nonchalance down to laziness or ignorance, but it seems much more likely that exceptionalism and fatalism play a large role in the way Americans assess not only the flu, but also health risks in general. The dangers of drugs and of unsafe sex, for example, are well documented, yet even in the hyper-informed West, people's responses to these dangers can often be unpredictable, to put it mildly. We shouldn't be surprised, then, that the MEP met similarly complex attitudes when its soldiers showed up with a plan to cure, three to five years down the road, a disease that people had lived with for generations.

All of which underscores again the human problems that make disease eradication so difficult. For the MEP, these problems did not end with the attack phase; on the contrary, the inability of the MEP to cope with the human side of the task was perhaps most apparent during the final phase of the program, consolidation. According to plan, the attack phase was intended to break the chain of transmission, but organizers understood that even in countries where this attack succeeded, plenty of vector mosquitoes would remain, as would plenty of parasites in the blood of infected people. Resurgence was inevitable unless the gains made during the attack phase could be consolidated, and this meant finding those people—malaria reservoirs, so to speak—and treating them. The actual eradication of malaria, in other words, occurred during the final, consolidation phase, with the aid of microscopes and antimalarial drugs.

The consolidation phase turned out to be the most difficult of the operation, and once again the difficulties had primarily to do with the people. As in the attack phase, adequate staffing was a problem. The work, which involved gathering blood samples on slides, was labor intensive and slow, and special skill was required in order to identify the parasites under a microscope. But worse than all this was the simple fact that many of the people who needed to be treated and monitored were asymptomatic: they had malaria but did not show it or sometimes even know it. Such cases, obviously, were extremely hard to identify, but they were harder still to treat. If it was increasingly difficult to convince people to open their homes to spray teams, imagine how difficult it was to get people who did not feel sick to submit to finger sticks and treatment. Yet all of the gains made during the attack phase depended on exactly that.

India offers a telling example of the human side of malaria eradication. With both an enormous population and an enormous malaria problem, India had been fighting malaria for decades before the WHO initiated its eradication program, and in 1946 had begun the largest control project ever attempted in the tropics. By the time the WHO made eradication its worldwide goal, in 1955, India already had a model antimalaria army in place. Each team consisted of 22 full-time men working 11 portable spray pumps. These spray teams worked under four supervisors, who in turn answered to a malaria inspector. The chain of command continued up to the highest level of the government.

When India made the switch from control to eradication, it was calculated that 390 of these teams could cover the entire country—a population of almost 400 million, or about 1 million per team. At the time, the organizers believed that eradication could be achieved if each sprayer, supervisor, inspector, and administrator did his or her job with the utmost discipline and precision. In an effort so vast, that was difficult to ensure. Nevertheless, the initial results of the control/eradication campaign were remarkable: between 1951 and 1961, malaria cases dropped from 75 million to 50,000.

It proved impossible to hold those gains, however, and human factors began to emerge in an increasingly important impediment. The MEP strategy required that eight out of ten houses be sprayed once every six months. Spraying left a white, chalky residue on the walls that smelled a little like chlorine. The wealthy didn't want their houses sprayed and so began bribing the inspectors to keep them away. DDT was very valuable, and some of it found its way out of the hands of the MEP workers and onto the black market. By 1965 there were 100,000 malaria cases, but by 1976, there were 6.4 million cases. Human nature, complacency, dwindling financial and political support, and a change in strategy from vector control to case finding and drug treatment were primarily responsible for the resurgence of malaria. Today, India records more than 2 million cases of malaria per year, although that number is still far less than the 75 million per year before the campaign took place.

Fatal Inflexibility: One Plan To Fit All

The battle plan for the MEP was a classic Soper operation. From Fred Soper's perspective, eradication had little to do with the mosquito or the people affected by the disease. Rather, it had everything to do with motivation, discipline, organization, and precision in execution. The MEP was the equivalent of a military operation. It had generals and weapons and foot soldiers. The big difference, however, was that the MEP had only one plan of attack. Regardless of the battleground, the strategy was essentially the same. The plan was spelled out in procedure manuals so highly detailed that even the charts to be displayed on the office walls of the administrators were dictated. The plan was not prepared to flex to take into account variations from region to region—either in the biology of the disease or in the social factors in each community. In fact, the plan had no provisions for field research, and, consequently, variations that existed or developed over the course of the MEP were never fully appreciated.

The MEP was strictly a top-down effort. It was not well integrated into the national or local health services of the countries affected, nor was it intended to be. It was a tenet of the WHO directorate that the eradication effort would fail within a given country unless it had support at the very highest level of government. Thus the program director of malaria eradication reported directly to a country's head of government rather than its health minister. MEP personnel were also entirely separate from the country's health authorities. They had no other duties, they were generally of higher caliber, and their pay was almost always higher than that of their colleagues. Because the number of personnel required for the MEP's plan was so high, MEP workers often outnumbered the personnel in all the other government health programs combined. All this predictably bred resentment and established barriers that impeded the flow of information that might have helped both the malaria campaign and other ongoing efforts.

The fundamental weakness in this rigidly defined, albeit meticulously executed, approach can best be appreciated by example. Wilbur Downs, a malaria researcher, had been working in Mexico for some six years prior to the MEP, and he has written about his memories of the beginning of the campaign. He was summoned to a meeting where Soper was presenting the plan for Mexico. As Downs listened, he recognized serious problems in the approach based upon his own experience. The habits of the principal vector mosquitoes, *A. pseudopunctipennis* and *A. albimanus*, were not well understood. The predominant local building material was adobe, and the clay absorbed DDT and accelerated its decomposition. When Downs spoke up about these issues, "Soper strode over to me, put his hands around my neck, and shook me vigorously, saying . . . it was this kind of talk which was impeding the malaria eradication effort around the world." The plan went forward and Downs was informed that his research in Mexico was superfluous and was to be terminated.

The MEP simply couldn't accommodate such variations even when they were known. That, coupled with the failure to conduct adequate field research throughout the campaign to monitor for such variations, ultimately contributed significantly to its failure.

The Price of Failure

The defeat of the MEP, according to one observer, seemed to "knock the stuffing out of the WHO staff." Funding was withdrawn or sent in other directions, and malaria control programs within endemic areas collapsed, leaving only the local health care systems to cope with the aftermath. Most of them were completely unprepared, in part because of the isolation of the MEP from the main systems of health care delivery. Perhaps equally distressing, funding for malaria research dropped dramatically. This translated into a dearth of new tools for combating the disease over the ensuing years, including an effective vaccine against the parasite. According to one malaria researcher, "Since they couldn't eradicate malaria, they eradicated the [malaria researchers.]" This left each country to battle the parasite largely on its own for almost three decades.

During that interval, the situation with malaria worsened. Widespread agricultural use of DDT generated increasing numbers of DDT-resistant mosquitoes. Rachel Carson's 1962 landmark publication of *Silent Spring* mobilized environmentalists with her chilling report of the effects of indiscriminate use of DDT—the death of songbirds and raptors was a warning that the powerful chemical had found its way into other important biological entities including fish, livestock, and house pets. Ultimately, DDT found its way into a high percentage of the world's human population as well, where it is still suspected of posing a threat to health. Environmental concerns were coupled with the less altruistic goals of the chemical companies, many of which wanted to sell their more profitable newer insecticides and lobbied governments accordingly. By the beginning of the 21st century, DDT had been banned in all but about two dozen countries, and a strong movement existed to ban its use entirely.

Although a ban on DDT was more than reasonable from the ecological perspective, its result on malaria control in countries that had the resources to continue the use of DDT was substantial. Not that it is the only insecticide; DDT is simply the most effective for controlling parasite-laden mosquitoes. A prime example is Belize, a small island in the Caribbean. Spraying rural homes with DDT came very close to eliminating the disease in the 1960s, and this success continued as long as DDT was in use. As the government tried to phase out DDT in the 1980s, however, the number of cases of malaria escalated alarmingly. The government returned to spraying, and malaria was brought back under control. In 1999, DDT finally had to be abandoned. Faced with a vastly diminishing market for the pesticide, the only remaining plant in the Western Hemisphere shut down. There was no backup source. Belize was forced to turn to an alternative, a pyrethroid insecticide called deltamethrin that was three to four times as expensive as DDT and arguably less effective. The cost of the chemical alone consumed almost 90% of Belize's malaria control budget, leaving little for treating patients, draining mosquito breeding grounds, or conducting surveillance. Similar stories were repeated in other countries where DDT had previously been successful in controlling malaria. Although the insecticide is still being used today in some countries such as South Africa, little is known about which vectors remain susceptible to its effects.

In the meantime, the parasite was playing out its own evolutionary game amid the antimalarial drugs. Resistance to chloroquine, still the most effective and least expensive of all the antimalarial drugs, began to emerge in 1960, right on schedule. First appearing in Southeast Asia, malaria parasites resistant to chloroquine now inhabit all the endemic countries. Chloroquine is essentially ineffective in Southeast Asia and South America against *P. falciparum,* and its effectiveness is rapidly diminishing in many parts of India and Africa. Resistance to chloroquine has recently appeared even among strains of the more benign *P. vivax.* This disconcerting adaptation by the parasites has been echoed with other inexpensive drugs, making treatment increasingly difficult and costly.

The inadequacy and/or failure of local public health delivery systems to control malaria contributed enormously to the emergence of drug resistance as well. Among the quickest ways to foster resistance to effective drugs is to give inadequate doses. This exposes large numbers of parasites to low concentrations of the drug—enough to allow parasite survival and selection of strains that are resistant to the drug's effects.

The more resistant strains replicate and get transmitted to new hosts, and their population expands. When the MEP was shelved, this is exactly what had begun to happen.

Several factors in the delivery of chloroquine to victims of disease fostered drug resistance. During the MEP, antimalarials were administered as a part of the program. Their quality was assured, and malaria workers in the field provided the proper doses. After MEP, there was no mechanism to guarantee the quality of the drugs, now obtained from a variety of private outlets, so supplies were often lacking in stated potency. Second, the resources for public funding dried up. This left the people who were too poor to purchase even chloroquine more likely to use inadequate doses to save on the cost of the drug—if they used it at all. Even when the victims could afford them, drugs were frequently self-administered using inadequate dosing schedules. Consequently, some areas of the world, such as the northwest part of Thailand, have essentially no effective, inexpensive antimalarial drugs to which the parasite is susceptible. Needless to say, this has changed the equation between human and parasite dramatically.

This is not to imply that the activities undertaken during the MEP were directly responsible for the emergence of resistance in mosquito and parasite. Quite the contrary: most scientists believe that the mosquitoes' resistance to pesticides—DDT in particular—came about primarily because of the massive use of pesticides in agriculture, not from the comparatively small amounts used to spray walls. Parasite resistance to antimalarial drugs was hastened not so much by use of the drugs as by their misuse when the MEP teams disappeared and communities were left to their own devices for managing fevers.

This does suggest, however, that the failure of the MEP to integrate into the local and regional delivery systems, thereby strengthening them, was an opportunity missed. This mistake took a major toll, but it was at least one mistake that has not been repeated in subsequent campaigns against this and other diseases.

Forward to the Present

It is July 2001, the beginning of the rainy season in Bandiagara. The roofs of the wattle and daub houses have begun to melt away and the dusty roads, gardens, and fields are becoming swamps and quagmires again. Over the next six months, the southern half of Mali, where Bandiagara is located, will turn into 600,000 square kilometers of mosquito breeding grounds. Although this is the jumping-off place for tourists traveling into the ancient Dogon villages along the Bandiagara Escarpment, no more than a single bus a week passes through the town. The Bandiagara townspeople have little or no contact with the outside world. The rainy season, however, brings with it one of their most important contacts—the team of doctors and scientists who come here to study malaria.

Among the mosquitoes that breed in the vast number of pools here is *Anopheles gambiae*. As the numbers of this voracious mosquito increase, so will the number of people infected with the malaria parasite. Over 400,000 of Mali's citizens will suffer from the symptoms of the disease during this rainy season. Almost 253,000 of the sufferers will be children younger than five years old, and each child is at risk for developing the most severe form of the disease, cerebral malaria. At least 5,000 of these children will, in fact, suffer this grave complication, and half of these can be predicted to die.

The malaria parasite here is the deadly *Plasmodium falciparum*, but Mali is among the more fortunate countries where this parasite circulates. The parasite's resistance to chloroquine—still a safe, highly effective, and inexpensive drug—remains comparatively low. If those afflicted can get chloroquine they can usually expect to get well. Mali manufactures its own chloroquine tablets, ensuring that the quality is good and making it more likely that the drug will get to those who need it. Although everyone must pay for his or her own medicine, the cost of a tablet of chloroquine is less than a penny (U.S.). Mali is, in fact, one of the more fortunate malaria-affected African countries for other reasons as well. The country is not in the midst of civil strife or border wars. Mali has a stable government committed to improving the health of its citizens. The country considers malaria its number one health problem. Its previous president, Alpha Oumar Konare, was eager to participate in the new WHO malaria control program called Roll Back Malaria, and the current president, Amadou Toumani Toure, appears to be following in his path.

The WHO did not undertake this new malaria program lightly. Although the failed MEP was called off in the late 1960s, the memory is still fresh in the minds of many. Over the intervening decades, malaria has continued to represent an extraordinary public health problem among developing countries in the tropics. So in 1998, after nearly a decade of study and discussion, Roll Back Malaria (RBM) was born.

The stated goals of this new program—to halve the number of deaths from malaria throughout the world by the year 2010 and to build a sustainable approach to maintaining those gains—are far more conservative than those of the MEP. The MEP was noted for its rigid, single-pronged attack against the mosquito and for the program's centralized management, operating in isolation from all other health care activities in every country. RBM is, in theory, defined by flexibility and decentralization.

The reality of Roll Back Malaria is sometimes distinct from the promise, however. Three years after its launch in Mali, a country with commitment at the highest level of government, most of the activity has remained centered on planning and consensus building. The health ministry and its partners have designed a strategy that they believe will match the country's needs—a strategy based on analysis and field studies conducted at the national, regional, and community levels. Thus research, rather than a standard formula, informed their decisions, and a National Malaria Control Program emerged that involves not only the central government, but also participants down to the community level.

Mali has also begun to gather the financial and human resources it will need to conduct this new program. The Ministry of Health has assembled its partners, drawing upon various agencies of the UN such as UNICEF, the World Bank, the WHO, and the UN Development Programme; other countries, including Germany, Belgium, and Switzerland; and many nongovernment organizations such as Groupe Pivot Santé et Population and Plan International. Its partners have pledged their support to help offset the cost of the program for the first five years. Optimism within the Mali Ministry of Health runs high despite the fact that there will likely be a funding gap in the $US 43 million budget that must be filled some way.

The Mali attack plan bears little resemblance to the MEP's monolithic approach. The strategy thus far is expected to be multipronged, employing a variety of components to reduce the impact of the disease; each prong is intended to build capability within Mali's struggling health care infrastructure and research community. Its components include projects for improving management of patients in all settings, from hospitals with their physicians to home care provided by community workers; projects for preventing malaria, from promoting the use and availability of insecticide-treated bed nets to prophylaxis for pregnant women; projects to monitor for, predict, and react to epidemics; and initiatives to develop strong, focused research programs employing Malian scientists and physicians alongside their Western counterparts.

The story that continues in Bandiagara typifies the extreme challenges that this new battle against malaria will face, however. In this small town of 12,000 in central Mali, most people live on less than $US 1 per day. The average life expectancy is 47. The incidence of malaria here is 150%, which means that just about everyone has malaria at least once a year. The attack rate is even higher for children. Electricity comes only from generators, and the recently improved medical center is a few buildings constructed out of cinderblocks. Many of the clinics here have no running water, and some of the windows lack screens. Across the street from the medical center is a series of stone igloos that house the training center and manufacturing facility for the traditional healers who deliver most of the health care in the community.

Sitting comfortably in between these two worlds is another set of buildings. This set houses the team of doctors and scientists who have been coming to Bandiagara for over ten years to study malaria—the "disease of the green season." Ogobara Doumbo, a malaria expert and physician who grew up near here, leads the team, along with his more youthful colleague from the University of Maryland, Christopher Plowe. Doumbo brings a long and distinguished career in malaria research at the University of Mali, where he directs the Malaria Research and Training Center; Plowe brings his NIH credentials and a decade-long enthusiasm for the research that he hopes will ultimately benefit public health.

Malaria research is their primary reason for coming here; they are following drug resistance in the parasite and laying the groundwork for vaccine trials. The team's impact on the health of the community's children over the years, however, has been noticeable and reflects the goals espoused by the new malaria control program. Plowe remembers when they first came to Bandiagara. "Health care there was pretty minimal before the research teams arrived. The majority of children suffering from cerebral malaria, or wabu as it is known here, were brought to the traditional healers." Doumbo more than anyone appreciated the implications of this practice for the researchers. For starters, the team would need to gain the cooperation of the community if it were going to convince them to enroll children in its study. Doumbo correctly suspected that the community had more faith in the traditional healers than in the West-trained physicians, and thus the researchers would have to get active support from both. Second, it was likely that the outcome of using herbal remedies to treat cerebral malaria was comparable to receiving no treatment at all. If the healers could be persuaded to bring children with wabu to the researchers for care, the outcome should be drastically improved.

Doumbo learned that five of the traditional healers specialized in cerebral malaria and that they kept meticulous records of their results. From July through September of that year, they had treated 218 cases of severe malaria. This wasn't even the peak of malaria season. During the same time, the health center had seen no more than 11 cases. Plowe recounts the story. "Although most of these cases were not diagnosed in the laboratory, we could be pretty certain that they were cerebral malaria. Their fatality rate was about 50%, which is about what you expect with untreated cerebral malaria. The results at the hospital were not any better, probably because of issues of supply and delayed treatment. The healers agreed that this was a terrible success rate. Ogo [Doumbo] convinced them to set up a case control study to treat the cases of cerebral malaria through the research clinic. The traditional healers often brought the children there themselves, and the research staff were careful to return the children to the healers when they had recovered. The case fatality ratio quickly dropped down into the single digits and that was well recognized by the community and the healers. After that, they were very welcoming to the research team."

Without this critical intervention, it is unlikely that the team could have gained the consent of the community to conduct its research. Of equal importance, team members improved the case management of the most fatal form of the disease they were there to investigate. Now every year when they return, they meet with the community leaders and the traditional healers to discuss plans and gain their consent. The early bonds forged with these groups have continued and ensure that the community members are willing to participate in the group's studies. Plowe reflects on this early success: "An important part of getting to work with the community is getting to the right people, gaining the trust of the healers, and then getting the trust of the people in the community. [Without the participation of our Malian colleagues,] we would have had a very difficult time. I'm sure we would have gone to the community leaders and the local doctors, but the local doctors were much less connected with and much less trusted by the community, and there's simply no way that outsiders like us would have known to contact the traditional healers. They really are the trusted figures. They are key. If they decided we were no good, that would be it."

The presence of the research team over the years has helped improve other aspects of health care in Bandiagara. For example, by switching children with cerebral malaria from quinine—an expensive and toxic drug that must be administered intravenously—to an oral medication such as Fansidar as soon as possible, the team found a more practical and sustainable approach to treatment. The research team examines each of the 450 children it enrolls in its study group once a week during the malaria season as well as anytime a child is sick. Consequently, this cohort benefits in general from improved care, but the team will also see anyone who comes to its clinic, whether they're enrolled in a study or not. The members of the research team have made a point of working closely with the Bandiagara physicians, returning patients to them after they have been diagnosed and begun treatment. This has no doubt also served to improve the local physicians' capabilities. Plowe and his colleagues believe this is an important legacy that they will leave behind if and when the team discontinues its efforts in Bandiagara.

The Malaria Research and Training Center (MRTC) that Doumbo heads is in many ways a model for capacity building, another goal of the new MCP. The MRTC began as a cooperative program between the University of Mali, the Institute of Parasitology in Rome, and the National Institutes of Health in the U.S. Over the years, it has grown into a highly successful and sophisticated operation, training African physicians in public health and researchers in state-of-the-art technology. The MRTC team that goes to Bandiagara, as an example, typically includes two or three senior Mali scientists, two or three American scientists, and ten mid-level Malian researchers. A physician member of the team remains at the research clinic year round.

Unfortunately, this medical largesse does not extend much beyond larger towns like Bandiagara and the capital city of Bamako. In a village of about 6,000 people near Bandiagara, there is one dispensary and no clinic. When members of the team made a rare visit to the village, they found the shelves of the dispensary bare. Most villages in Mali don't even have dispensaries. The distances that must be traveled to get to a clinic are long, and the roads are poor. Although some make their way to towns like Bandiagara, Plowe suspects that many children simply die in their villages without medical intervention.

Plowe believes that the National Malaria Control Program (MCP) will take advantage of the enormous resources that the MRTC has to offer. "There was an outbreak of malaria in the north in an area that just wasn't safe to drive through because of the banditry in the area. The Ministry of Health wanted advice on how to deal with this outbreak, and so they flew a few members of Ogo's [Doumbo] team up in a military transfer jet. They couldn't stay long enough to do clinical efficacy [studies]. They collected blood by finger sticks, did microscopy on site, brought blood back on filter papers, and ran PCR on them in Bamako. They found that about 70% were infected with a parasite that had a mutation making it resistant to chloroquine, but not the mutation that causes resistance to Fansidar, so they were able to recommend using Fansidar to deal with the epidemic in real time." Interactions like these will be critical for RBM and the MCP.

Prevention, another goal of the new malaria control program, is another matter. Plowe is unaware of any concerted efforts under way in Bandiagara directed at prenatal prophylaxis. The members of the community, including the research team, haven't been inclined to adopt use of insecticide-impregnated bed nets, a major WHO and now MCP strategy. The nights are hot, and bed nets are simply not a part of the culture. Abdoulaye Djimde, a Malian graduate student working with Plowe at the University of Maryland, lived in Bandiagara as a child, and his view helps put the issue into perspective. "In places [like Bandiagara] where people are not used to sleeping under the nets, it is really very hard to make that change. Even if you bring in the nets and make them available, you have a very intensive educational effort to undertake—getting mothers to take the babies under the net as soon as the sun goes down, getting mothers to join the babies when the mosquitoes start biting. If you have the net inside the house and you stay outside for three to four hours in the night, you have enough time to get your share of biting before you get under the net . . . if you want it to be something that goes into people's way of living, it takes much more time and effort."

Many who are familiar with the problems faced by the RBM remain quietly skeptical and fear that even this limited effort is not realistic. The tools simply aren't available and the resources aren't sufficient to effect much change. Plowe doesn't believe that the malaria problem will be solved until economic development begins to transform Africa. "Perhaps RBM can assist that, but it is a bottomless pit. It's a lot to ask a country with limited resources to invest heavily in something that is only palliative. Perhaps when the right tools are available, it will make more sense."

Still, each small step makes a difference. Djimde reflects, "Right now the best thing we have is case management, so either improving diagnosis or good algorithms for clinical diagnosis to get treatment in place quickly, are important. We make sure that the community has direct benefit from the research we do. . . . The community really looks forward to our coming back during the rainy season. I can tell you that if we are late in coming, they send messengers to ask 'Are you coming this year? When are you going to start?' And they see the benefit, even for some of the issues that we are not following. Since we began coming here, they see a sharp drop in infant mortality. They tell us 'Our kids are much healthier since you have been coming here.' "

Magic Bullets

Despite its ultimate failure to eradicate malaria, the MEP's efforts helped reduce the burden of disease in many parts of the world during its 14-year history. In the subtropical climates of southern Europe, in the island settings of

Mauritius and Singapore, in Hong Kong, and in parts of Malaysia, malaria was either eliminated or brought substantially under control, saving millions of lives. While all the gains were not sustainable, there were positive benefits from the program over the short term, even in countries such as India and Sri Lanka, where malaria rates returned to their previous levels.

The premises upon which hopes for eradication were based remain credible even in the aftermath of failure. Attacking the mosquito reduces the number of bites, which decreases the number of infective doses delivered and subsequently the number of infected people to pass along the disease. Attacking the parasite decreases the number of people with the parasite in their blood, which decreases the number of infected mosquitoes, which decreases the number of infective doses delivered. Sustaining one or both of these attacks for a long enough period of time should allow the disease to burn itself out, just as postulated some 50 years ago.

The challenges inherent in tackling a disease that is rooted so firmly in our ecosystem remain daunting, however. The tools available to those who tried to eradicate malaria—insecticides and antimalarial drugs—were not sufficient, even if they were applied in combination. Surface water management, better materials for construction of housing, and access to adequate health care were outside the realm of economic feasibility for the developing world. These factors are as true today as they were in 1969.

The failure of the MEP took its toll in many ways, but among the worst was the scientific hiatus created by the effort, which compromised the discovery process for the next three decades, leaving the latest campaign with few new tools to fight the coming battle. Investment in malaria research slowed to a trickle during the MEP, a consequence of the overstated promises of disease eradication. After the MEP, economic incentives that might have driven research and development didn't exist. Malaria had become almost exclusively a disease of the economically disadvantaged.

With the increased flow of investments and an aggressive research agenda such as that of the MRTC in Mali, the pipeline of discoveries is beginning to fill again. Most public health specialists believe that the most promising route to preventing malaria, and a route that could potentially return us to the notion of eradication, lies in the development of a reliable vaccine—a development that is unlikely to be realized by 2010, the deadline for this newest WHO initiative.

Before the MEP shut down, another war against a disease had begun. This war did have an effective vaccine as its weapon. That, coupled with lessons from the MEP, created a story with a very different ending.

Malaria, Man, and Mosquito: the Biologic Perspective

Man and Malaria

Two of the malaria parasites, *Plasmodium falciparum* and *Plasmodium vivax,* are far more common in human infections than other plasmodia. Both these parasites are complex in their own right, but their interactions with people are particularly intricate. *P. falciparum* is the more fearsome of the two. Restricted primarily to tropical climates, it accounts for the most devastating form of the disease.

There are two primary reasons for this. The first is the ability of *P. falciparum* to infect red blood cells at every stage of cell development—meaning, in other words, that *every* red blood cell, old or young, is a target. In each infected cell, the parasite multiplies and produces between 24 and 32 new parasites, which then infect new cells. Over a relatively short period, half or more of an individual's red blood cells may be taken over by the parasite. At the end of each replication cycle, the newly minted parasites burst the red cells, leaving the victim anemic, fevered, and ill.

The second reason for the devastating effect of *P. falciparum* is the change this parasite causes in red blood cells *before* they burst. Infected cells develop "knobs" on their surface—abnormal structures that make the cells stick to the lining of blood vessels. The consequences of this can be catastrophic for the liver, kidney, and other organs—the brain in particular—because the tiny capillaries that feed these organs become clogged with knobby, parasite-stuffed cells. This restricts the flow of blood and thus of oxygen, in essence starving the organ. This starvation underlies the most severe and often fatal form of the disease, cerebral malaria.

P. vivax, on the other hand, is far less likely to cause serious damage. In contrast to *P. falciparum,* it infects only young red blood cells. This limits the total number of potential harbors for the parasite to only 5% or less of the circulating cells. Thus, when *P. vivax* bursts the red cells, the damage is less extensive, the anemia is milder, and the symptoms are less severe. More importantly, the infected cells don't develop knobs and don't stick to the inside of blood vessels. *P. vivax,* however, has a trick that isn't available to *P. falciparum.* The parasite can remain dormant in cells in the liver for weeks or months and then suddenly activate, precipitating a relapse of the disease and providing an active malaria reservoir for a passing mosquito.

Malaria and the Mosquito

Malaria is spread by at least 50 species of mosquito. All belong to the genus *Anopheles.* Each has its own habits and preferences, any of which can become a factor in the spread or control of the disease. Some mosquitoes prefer to feed on humans, and some prefer animals. Some breed in the shadows, some in sunlit pools, and some in tin cans and discarded automobile tires. Some prefer large ponds; others, shallow puddles. Some feed at night; some during the day. Some live in the tropics, and others live in temperate climates. Some are highly efficient in spreading malaria to humans, and others are much less so.

The worst mixture, by far, of human-biting mosquitoes and malarial parasites is found in the tropical climate of sub-Saharan Africa. Here and only here do *P. falciparum* and a mosquito named *Anopheles gambiae* coexist. *A. gambiae* is a voracious insect: simply by exposing their legs to be bitten, volunteers working with research stations in this region can trap up to 300 mosquitoes in eight hours. As if this were not enough, the warm tropical climate favors not only the mosquito, providing ample breeding grounds, but also the parasite maturing in the mosquito's gut, shortening the time the parasite needs in order to develop into the stage that infects humans. In such a climate, the combination of *P. falciparum* and *A. gambiae*—the worst malaria parasite and the most efficient transmitter of malaria parasites—produces the highest infective rate in the world. It's possible for one person in sub-Saharan Africa to receive as many as 1,000 infective doses of malaria every year. Given these conditions, eradicating malaria from this part of the world is considered highly unlikely, if not impossible.

Humans Evolving

The parasites that cause malaria seem to have spent only a short period with their human hosts in evolutionary time, perhaps emerging as recently as 4,000 to 10,000 years ago when alterations in climate and human behavior brought about favorable conditions. The fiercest of these microbes, *Plasmodium falciparum,* accounts for most of the 300 million cases of malaria and the million or so deaths that result each year.

Among those people most at risk, however, exists a curious set of genetic traits that confers resistance to the worst effects of the disease. These genetic traits are the result of mutations that affect properties of the red blood cells, the site of the parasite's attack. The appearance of these traits in human populations traces back only some 5,000 to 10,000 years, a time coincident with a massive expansion in the parasite's population. At this same time, a climate shift in West Africa changed the Sahara from an arid wasteland into a green savanna with many lakes and ponds—perfect breeding grounds for the mosquito. Africans, who were still fishing and herding, began to live in denser communities along the lakeshores, providing an ever-increasing opportunity for the parasites to propagate. Then, between 2,500 and 4,000 years ago, Africans began to clear forest for agrarian purposes (so called slash-and-burn agriculture). Their populations further increased, and their land use created more sunlit pools, ideal conditions for both mosquito and malaria.

One of these human genetic mutations occurs in the gene that controls the shape of hemoglobin, the molecule that carries oxygen in red blood cells. The mutation produces an abnormal form of hemoglobin, giving rise to sickle cell disease. A second mutation produces an abnormal form of glucose-6-phosphate dehydrogenase, another important molecule in red blood cells. Both genetic traits are, on the surface, disadvantageous to humans: both result in anemia, and sickle cell disease is potentially fatal.

Such mutations are normally eliminated from the gene pool because of their negative effects. However, both offer a selective advantage in surviving malaria, making the red blood cell far less hospitable for the parasite. And those most at risk for infection—the people of sub-Saharan Africa—commonly carry these traits. Scientists have long shared the view that the maintenance of these adverse genetic traits demonstrates how powerful the impact of malaria has been on human evolution.

A New Magic Bullet

No one today talks seriously about malaria eradication, but given the difficulties with available control strategies—biologic and social resistance to insecticides such as DDT, parasite resistance to antimalarials, and cultural resistance to impregnated bed nets—many health officials have begun to feel that even malaria control will depend on the development of a practicable vaccine. The hope that a vaccine against malaria can be developed is bound up in the observations of what happens during normally acquired infections. Malaria is worst in those with little or no experience of the disease—adults from outside the endemic region and young children. Once a person has suffered through several bouts of malaria, however, partial immunity begins to develop; partial because the parasite still proliferates in the person's bloodstream, but its damaging effects are mostly suppressed. Thus adults living in endemic areas such as sub-Saharan Africa are commonly infected but suffer few or no symptoms of the disease. They have naturally acquired immunity.

Of course, a look at naturally acquired immunity also discloses a set of biologic realities that make the development of a vaccine an extraordinarily challenging task. Naturally acquired immunity does nothing to protect against infection. All it does is suppress the parasite's activities to a level that is less evident to the person infected—in other words, the symptoms are less noticeable and the risk of serious disease is lowered. While this might be a reasonable outcome, all things considered, it hardly provides an avenue through which the disease can be controlled.

Other biologic realities are even more daunting. Although humans and the malaria parasite may have only coevolved over the past several thousand years, the parasite is a wily foe. To begin with, it is a large and complex organism. Unlike a simple virus, with its monomaniacal attack strategy and limited number of component parts, the malaria parasite has somewhere between 5,000 and 7,000 genes that code for proteins, any of which might be involved in some way in the parasite's ability to invade. The parasite also goes through a complicated life cycle, changing its shape and its surface with each stage and presenting its host—and vaccine researchers—with an array of immunologic disguises. The blood-borne stage of the parasite, the merozoite, even changes surface components as the immune system recognizes them.

Complicating all of this is the fact that the parasite spends very little time free in the bloodstream, where it is most vulnerable to attack by the immune system. Only a few minutes after the mosquito injects them, the sporozoites begin to invade the cells of the liver, where they are protected. There they begin an astounding multiplication, with each invader generating up to 30,000 merozoites. Each merozoite will ultimately leave the liver and, in a matter of seconds, infect a red blood cell, where it is once again sheltered. This ability to hide, multiply, and then suddenly attack—sort of a biologic version of guerrilla warfare—presents vaccine researchers with an especially difficult task. To miss even one sporozoite is to leave open the possibility of infection; and multiple bites, which are the rule in areas of high endemicity, mean that the supply of sporozoites is constantly renewed. So far, the malaria parasite has been a moving target that even the steadiest vaccine researchers have been unable to hit consistently.

Despite this rather intimidating array of problems, researchers began searching for a vaccine in the 1940s. The first experimental evidence that a vaccine might be possible arose from an odd approach. Infected mosquitoes were zapped with radiation and then allowed to bite a volunteer. The sporozoites they injected were robust enough to cause an infection but too debilitated to produce symptoms of disease. Once infected, the volunteer was protected against malaria for as long as nine months. Of course, this hardly offered a practical approach to a vaccine, but it spawned a furious competition among several laboratories. None was able to deliver a usable vaccine, but each helped develop a better understanding of what it takes to make one. A better understanding, however, hardly generated a significant amount of optimism that a vaccine could be developed easily, and with the promise of the Malaria Eradication Programme, funding essentially dried up.

Today dollars are flowing back into malaria research, and some of that funding is driving renewed efforts to develop a successful vaccine. It will likely be a vaccine with multiple components, targeted at different stages of the parasite's complex life cycle and the different proteins of those stages, including the critical reproductive stage in the mosquito's gut. It may be an outgrowth of a new vaccine technology that uses the DNA coding for the various components directly rather than the components themselves. Even with renewed interest and renewed optimism, however, the malaria parasite remains an elusive foe, one against which even the most optimistic won't forecast success in less than a decade.

Smallpox: the Right Disease, the Right Time

They do die in these parts . . . of the small pokkes and mezils.

PACE (1518)

Smallpox no longer afflicts humanity, and the remaining virus will be confined to a few laboratories under high security.

DONALD A. HENDERSON (1978)

October 12, 1977, was a fateful day for Ali Maow Maalin, although it probably didn't seem so at the time, and an important day in the history of smallpox. On that day, a driver searching for the smallpox isolation camp near Merca, in Somalia, stopped to ask for directions at the local hospital where Maalin worked. The young man offered to show the driver the way to the office of the director of the local smallpox surveillance team and jumped into the car. Two small children sat in the back seat.

The virus that causes smallpox infects only humans. It has to pass from person to person in an unbroken chain to maintain itself. Someone breathes in the virus, becomes infected, and then breathes it out again. By 1967, when the WHO began its Intensified Smallpox Eradication Programme, this chain was approximately 3,000 years long. During those millennia not one day passed when the virus was not reproducing in someone somewhere.

Ten years after the program began, in October 1977, on the Horn of Africa, that chain was about to be broken. Somalia was the only place on earth where smallpox transmission still occurred; the disease had been stopped everywhere else. Consequently, Somalia was the focus of intense activity and international scrutiny. Extra money had poured in, along with vehicles and personnel, and for months eradication teams had been tracking down those infected with the virus and vaccinating everyone around them.

By far the most difficult area was the Ogaden desert, which spans the border between Somalia and Ethiopia. The usual smallpox eradication strategy of surveillance, isolation, and targeted vaccination had worked well in most places, but it was not well suited to populations on the move like the nomads of the Ogaden. It was often almost impossible to know who had the disease or where, from day to day, such persons might be. Fewer than 20% had been vaccinated.

When infected individuals were discovered, they were moved as quickly as possible to an isolation camp. Until isolated, every infected person threatened every unprotected person who came near, however briefly. The two children in the back seat of the car were among the infected, and Ali Maow Maalin, as it turned out, had no protection against the virus. The ride from the hospital to the director's office was less than a kilometer, but for Maalin that was far enough.

The virus silently invaded cells in Maalin's throat and began to replicate, making more and more copies of itself. Soon, the virus moved to lymph nodes, where it continued to multiply—each virus particle becoming 10,000 new ones in the space of a few hours—and from there into the bloodstream. On October 22, ten days after his fateful ride, the virus was no longer silent. Maalin developed a fever. His head hurt and he began to vomit.

When he was admitted to the hospital on October 25, no one suspected smallpox. Maalin had been vaccinated against smallpox, after all—though, as was often the case in poorer countries, not effectively. He had even worked as a vaccinator in the eradication program; there was no reason to think he might have smallpox himself. By October 26, the virus had made its way to his skin, invading the cells there and generating the skin eruptions that are the hallmark of smallpox. Still no one suspected the true nature of his disease, perhaps because Maalin had the relatively mild form of smallpox called variola minor. His physicians changed their diagnosis from malaria to chickenpox and sent him home.

Soon the features of Maalin's infection made its agent undeniable. The skin eruptions had filled with pus. They covered his body from head to foot, first weeping and then scabbing over. Maalin must have known that he had the disease, as finally did the nurse who reported him to health officials.

When the smallpox eradication staff confirmed that Maalin had smallpox, he was quarantined in his home under 24-hour guard and then transported to the same isolation camp as the children who had exposed him. The hospital was sealed under 24-hour guard. All the patients and staff were vaccinated, as were all the people in the 792 houses in the area where Maalin lived. Officials searched the entire town for six weeks, setting up checkpoints on the roads and footpaths so that anyone coming or going could be vaccinated. According to the official WHO history of the global eradication program, 54,777 people were vaccinated in two weeks because of Ali Maow Maalin.

Of the two children in the car, one recovered and one died—she became part of the unlucky 1% who dies from variola minor. By the end of November, Maalin himself had recovered. Before he was correctly diagnosed, he had exposed 161 other people. None of them developed smallpox, and that left Ali Maow Maalin to enter the history books as the last case of naturally acquired smallpox in the world.

Fading Scars

Although far less common than malaria, smallpox was a more frightful disease. A Western reader might see malaria primarily as an inconvenience, as one of the risks of traveling, even as part of the adventure of globetrotting—a few mosquito bites and some fever, a rigorous course of antimalarial drugs, and afterward you have an exotic story to tell. For a certain cohort of college-age travelers from the U.S. and Europe, contracting malaria while backpacking through the developing world is almost a rite of passage. Smallpox, however, always presented an uglier face, one not to be trifled with: swollen pustules, scars, blindness, death. Up to 30% of people afflicted with variola major, the worst form of smallpox, died.

Smallpox was also, as it turned out, a more vulnerable disease than malaria. Many of the people reading this book have smallpox vaccination scars on their upper left arms, but their children may not, and their grandchildren almost certainly will not. Those scars, like the disease that made them necessary, have faded from our collective memory. Most physicians practicing today have never seen a patient with the disease, for since Ali Maow Maalin's case, no naturally occurring cases of smallpox have occurred anywhere in the world.

For centuries smallpox was the source of mystery and indiscriminate terror, for the great and the small alike. Paleopathologists inspecting the mummy of Ramses V speculate that he died of smallpox in 1157 B.C.E. Queen Elizabeth I of England was diagnosed with smallpox in 1562, and Queen Mary II, of the English royal couple William and Mary, died of it in 1694. Abraham Lincoln contracted smallpox in 1863. The famous names only underscore the point that smallpox was a disease no one could avoid; the bulk of the cases over the centuries, of course, fell on those without famous names, and millions fell prey to it. Even as late as 1967, at the beginning of the final push to eradicate the disease, 10 to 15 million cases were still occurring annually, with as many as two million of the cases ending in death. And once the disease set in, nothing could be done except to watch and await its outcome.

The victory over smallpox is arguably the single greatest achievement in the history of medicine. Donald A. Henderson, who directed the WHO eradication campaign, called smallpox "the most devastating and feared of the great pestilences." Its eradication is memorable on those grounds alone. As an example of international cooperation and effectiveness, however, the Smallpox Eradication Programme also belongs in a wider context: the ongoing efforts at and debates over disease eradication. Especially when viewed alongside the Malaria Eradication Programme, with which it overlapped by several years, the SEP illustrates many of the differences between a campaign that succeeds and one that fails, and current eradication and control efforts have all been shaped by those differences. Ironically, it is this same achievement that has left the world 25 years later susceptible to a virus that was never destroyed, for while the disease was eradicated, the virus itself never was.

From Golden Needles to Vaccine

Humans had lived with smallpox for thousands of years, and many approaches to treating the disease had developed over the centuries. Some physicians recommended lancing the pustules with a golden needle. Some dosed their patients with sugar and barley water, some with sulphur and barley water, some with gold mixed with mineral water, and some with horse dung or sheep dung. There were opiates and emetics. Bleeding was popular, of course. Praying worked as well as anything did, so people tried that, too. In China, Ch'uan Hsing Hua Chieh was the goddess of smallpox. For Hindus, her name was Misanima. Christians could solicit St. Nicaise, St. Sebastian, St. Roche, and St. Barbara.

Perhaps the longest-lived of these treatments was erythrotherapy, the notion that the color red was therapeutic for smallpox. John of Gaddensden wrote in 1335, "When the son of the renowned king of England lay sick of the smallpox, I took care that every thing around the bed should be of the red colour; which succeeded so completely that the prince was restored to perfect health, without a vestige of a pustule remaining." Even by the early 20th century, some U.S. hospitals offered red rooms—red glass, red curtains, red walls, even red globes around the artificial lights—in which patients could be treated. The chief proponent of erythrotherapy, the Danish physician Dr. Niels Finsen, wrote in 1903 that the "action of light on the course of smallpox is astonishing, and the effect of the red light treatment is one of the most striking results known in medicine."

Of course, neither erythrotherapy nor any of the other remedies worked. Smallpox is an untreatable disease. Smallpox could, however, be prevented, and from this knowledge the tool for eradicating it was ultimately developed. People had known for centuries that it was possible to avoid the horse dung, the emetics, and the red curtains by avoiding smallpox altogether through a process called inoculation. Inoculation rests on the courageous idea that it's better to charge the enemy—in this case, smallpox—than to hang back and wait for a full-scale attack. By purposefully contracting the disease, people were often able to limit its severity. They became sick, but usually not gravely, and having passed through the fire more or less unharmed, they were safe.

The Chinese had carried out inoculation for centuries, using pulverized scabs from mild cases of the disease; they took the dust up their noses. In parts of 17th-century Europe, inoculation was a folk practice known as "buying the smallpox." In Turkey, the procedure was almost recognizable by modern standards: a practitioner scratched open a vein and introduced a tiny amount of smallpox. Lady Mary Wortley Montagu, who traveled to Constantinople in 1716, observed the practice there, had it performed on her son, and worked to popularize the treatment when she returned to England in 1718.

Her promotion, along with that of others, worked. Inoculation was accepted relatively quickly in England—George I set an example when he allowed his granddaughters to be inoculated—and it was well established by the middle of the 18th century. The College of Physicians endorsed the practice in 1755, and after the 1760s more and more people in England were inoculated free of charge. For many towns and villages, it made more economic sense to inoculate the poor for free than to care for them during an epidemic. Two hundred years later, as the WHO was beginning its eradication campaign, similar economic considerations were still relevant.

Wide acceptance was much slower in the rest of Europe. People harbored religious fears that circumventing smallpox, if in fact inoculation could accomplish this, might be circumventing the will of God. Another source of resistance was more mundane: it appears that for decades the French thought inoculation was a bad idea because the English thought it was a good one. Other concerns were better grounded in scientific reasoning. Inoculation did present genuine dangers, both to the person inoculated and to those near him afterward, and no proof yet existed that taking these risks conferred lasting benefits. Could a person get smallpox twice? No one was certain of the answer. They were certain of two things, however: inoculation could cause serious disease in some people, and the virus could be transmitted after inoculation to others through close contact. Some concerns proved to be justified; the practice of inoculation was one of the reasons centuries later that the smallpox virus continued to circulate during the last months of the eradication effort.

The prevalence of inoculation in England and the scientific interest in it led to the first vaccine in medical history. Edward Jenner was an English physician. He had been inoculated himself as an eight-year-old, and in the 1760s he began to ponder the connection between smallpox and another disease called cowpox. He had learned anecdotally from his patients in farm country that milkmaids who contracted cowpox, a milder disease that usually infects cows, were resistant to subsequent smallpox infections. Others were aware of the connection, but it was Jenner who first tested this idea experimentally. In 1796, he took some cowpox from an infected milkmaid and used it to inoculate a young farm boy. The boy had the anticipated slight reaction to the disease. Six weeks later, Jenner inoculated him with smallpox, but nothing happened; the second virus was innocuous because the boy had become immune.

This was the first proof that cowpox was effective in preventing smallpox, and it presented the world with a safe alternative to inoculation. The two viruses are so similar that the immune system can't distinguish between them. Having learned to identify and attack cowpox, the immune system is ready to do the same to smallpox. Because of Jenner's work, people were able to avoid smallpox altogether, even the mild form that, in the best cases, resulted from inoculation.

Accustomed as we are today to state-of-the-art Western medicine, it seems almost funny that the first vaccine was not a multimillion-dollar wonder drug developed in a laboratory, but a drop of pus taken from a sore on the hand of a milkmaid. The word *vaccine* itself is only an Anglicized form of the Latin word for *cow,* and its medical use derives from the Latin name (coined by Jenner) for cowpox: *variolae vaccinae,* smallpox of the cow. As mundane as its beginnings were, smallpox vaccine and its ability to prevent the disease became a significant factor when the notion of eradication began to emerge. The availability and effectiveness of the vaccine also came to represent a significant difference between the tools of the malaria and smallpox eradication efforts.

Moving Toward Control

Vaccination established itself much more quickly than had inoculation, particularly in the U.S. and Europe. By 1821, vaccination was compulsory in Bavaria, Denmark, Norway, Russia, Sweden, and Hanover, and the incidence of the disease in these countries began to fall rapidly. Continuing research and innovation across the continent solved some of the problems associated with Jenner's basic method, making vaccination both more effective and more available. Over time, for example, it had become clear that people vaccinated as children could still contract smallpox as adults, albeit a mild form. The solution was simple: revaccination. Where revaccination was made compulsory, as in the Prussian army during the 1830s, the results were dramatic. The incidence of disease plummeted. A second problem had been the availability of vaccine, and this was solved in 1843 when an Italian researcher began to inoculate cattle with the cowpox virus, in essence creating an entire herd of vaccine producers. It was then possible to lead such a cow from house to house, offering fresh, wholesome virus to all who needed it.

In spite of these advances, however, surprising numbers of people opposed vaccination, and smallpox remained a dangerous presence. The worst outbreak of the 19th century occurred in Europe between 1870 and 1875, when perhaps as many as half a million people died. Some still continued to argue that smallpox was a visitation from God and therefore had to be endured. Others framed the argument in more earthly terms. In 1826, Thomas Malthus, the political economist who emphasized the disastrous effects of unchecked population growth, wrote that "above all, we should reprobate specific remedies for ravaging diseases, and those benevolent, but much mistaken men, who have thought they were doing a service to mankind by projecting schemes for the total extirpation of particular disorders." According to this argument, smallpox was only one of many diseases that helped reduce the number of mouths to feed.

Another reason for the continued presence of smallpox was, ironically, the very success of the fight against it: as the disease receded, people began to take it less seriously. This was particularly true in England, where a substantial number of citizens worried less about smallpox than about the encroaching powers of government. Vaccination was free in England, but it was also compulsory after 1853, and many people resented the idea that the government could compel them to do anything. Although almost 16,000 people died of smallpox in London alone between 1870 and 1880, opposition to compulsory vaccination continued. In 1885, tens of thousands of demonstrators gathered in Leicester to rally against the vaccination laws. Well into the 20th century, English parents could claim reasons of conscience and refuse to have their children vaccinated.

Nevertheless the trend in England—as in the rest of Europe—was toward vaccination, and as the century turned the disease was retreating. In Sweden, for example, there were no cases of smallpox in 1896, 100 years after Jenner's experiments. World War I led to a surge of cases in many European countries, but afterward the retreat began again. Not only did the number of cases fall dramatically—Austria became smallpox-free in 1924, Great Britain in 1935—but the severity of the disease decreased as well. The worst form of the smallpox, variola major, had all but disappeared, and the cases that remained tended to be of variola minor—still a dangerous disease, but much less so. In North America, variola minor had been the predominant form for decades, and by 1920 the U.S. was recording only one death from smallpox per million persons. Even in the Soviet Union, which had recorded almost 200,000 cases of smallpox in 1919, the situation began to improve in the early 1920s with compulsory vaccination laws. By the end of World War II, endemic smallpox in Europe was limited to Spain and Portugal, and by the early 1950s, it had ended there, too.

In Africa, Asia, and South America, however, the situation was different. In India, almost 800,000 people died from smallpox between 1928 and 1937. In Afghanistan, perhaps as many as 300,000 cases of variola major occurred during some years. In Bangladesh, there were 58,000 *reported* deaths in 1958, and the actual number was probably at least ten times higher. When the WHO passed its first resolution calling for the global eradication of smallpox in 1959, either variola major or minor was endemic in 59 countries and territories. The population of these regions totaled over 1.7 billion, or roughly 59% of the world's total.

Thinking about Eradication

The idea that we might be able to eradicate smallpox can actually be traced back to Jenner. In 1801, he wrote, "It now becomes too manifest to admit of controversy, that the annihilation of the Small Pox, the most dreadful scourge of the human species, must be the final result of this practice." It was another 150 years before this idea surfaced in a serious way.

In 1949, the Pan American Sanitary Organization (PASO) proposed to attempt smallpox eradication in the Western Hemisphere, but WHO wasn't interested. Given the prevalence of smallpox, its frequently serious consequences, and its proven vulnerability to Jenner's vaccine, it's a little surprising to learn that the world community was not particularly enthusiastic.

Despite the WHO's lack of interest, PASO went ahead, but the effort was chronically underfunded. Although some progress was made in testing and producing freeze-dried vaccine, the program went essentially nowhere, and by 1958 it appears to have been forgotten—not only by the delegates to the World Health Assembly but also by the member nations of the PASO itself.

Two related reasons help to account for this lackluster reaction to smallpox eradication. First, though prevalent throughout South America, smallpox there was predominately the milder variola minor form. It was thus not the most pressing health problem in the region. Second, in 1949 smallpox was not the most pressing health problem anywhere in the world. Except for imported cases, it was virtually unknown in North America and Europe, including the USSR. Smallpox, in other words, was a regional issue, at least as far as the newly founded WHO was concerned—and a relatively minor regional issue at that. If the PASO wanted to launch a campaign against smallpox, it could do so, but the WHO wasn't inclined to contribute either money or support.

In fact, at mid-century, medical and scientific experts were by no means sure that eradication of a disease was even possible. Eradication efforts had worked fairly well against certain animal diseases—infected herds, after all, could be strictly quarantined or, if necessary, destroyed. Many knowledgeable people, however, felt that eradication of a human disease was beyond reach, and nothing about the malaria program was changing their minds. As late as 1965, René Dubos, a highly regarded scientist and spokesperson, was arguing vigorously in his book *Man Adapting* that the eradication of an infectious disease was a utopian concept, one we had neither the knowledge nor the resources to pursue. "Eradication programs will eventually become a curiosity item on library shelves, just as have all social utopias . . . eradication of microbial disease is a will-o'-the-wisp; pursuing it leads into a morass of hazy biological concepts and half truths."

Added to this skepticism about eradication was a certain amount of naïveté: not even the wealthiest nations of the world had much experience thinking about health in global terms. Like the WHO itself, the notion of an international health community was relatively new, and the organization's leaders were still trying to figure out what it meant to have a world health organization. What should be its priorities? How should it pursue them?

Smallpox clearly fell within the purview of the WHO, but the organization just as clearly did not know how to proceed against it. The World Health Assembly discussed smallpox in 1950, 1953, 1954, and 1955; the director-general of the WHO even proposed eradication himself in 1953. That proposal faltered, however—it was deemed "unrealistic"—and for the next several years, according to the WHO itself, "the issue of smallpox eradication lay dormant."

It was in this evolving context that the Soviet Union, at the World Health Assembly of 1958, raised the idea of smallpox eradication again. This resolution, too, might have failed had the Soviet epidemiologist Viktor Zhadanov not made so clear the need for global eradication, emphasizing the constant threat of imported cases that endemic countries posed to nonendemic ones.

The Soviet resolution suggested, in the roundabout language of a formal document, that when it came to infectious diseases, the world was fairly small: "In many regions of the world there exist endemic foci of this disease constituting a permanent threat of its propagation and consequently menacing the life and health of the population." In other words, the disease didn't respect national or regional borders. As long as people were crossing those borders, from nonendemic to endemic zones and back again, there was a permanent threat that smallpox would cross with them.

The Soviet resolution made the argument that smallpox was a global concern even more convincing by drawing attention to the money nonendemic countries were spending, year after year, protecting themselves against the threat of imported cases: "The funds devoted to the control of and vaccination against smallpox throughout the world exceed those necessary for the eradication of smallpox in its endemic foci." Though eradicating smallpox would cost a lot of money, *not* eradicating it was already costing a lot of money. Between monitoring borders, producing vaccine, administering vaccinations, treating adverse reactions to vaccinations, and mobilizing against periodic outbreaks or suspected outbreaks, the nonendemic world was spending hundreds of millions of dollars each year. The U.S. alone was spending $15 million on its efforts. Once smallpox had been eradicated, a substantial amount of money around the globe would be saved and could potentially begin to flow toward other health concerns.

Zhadanov's arguments proved persuasive, and the WHA adopted the Soviet resolution, which called for a careful study of the "financial, administrative and technical implications" of an eradication program. That study led in 1959 to another resolution, this one constituting a definite commitment to global eradication. Asserting that four or five years were sufficient for reaching the goal, the resolution called on the health administrations in endemic countries to set up eradication programs. The director-general was requested to assist these programs and to keep careful records of their progress. The resolution passed unanimously.

That unanimity is somewhat misleading, however. During the early years of the Smallpox Eradication Programme, 1959–1966, the campaign was simply not a high priority for the WHO. The organization backed the *idea* of smallpox eradication, and it offered encouragement, good wishes, and token support to any country attempting it. What was needed, though, was political will and money, and in the late 1950s and early 1960s both were going elsewhere. In 1959, malaria eradication was in its fifth year. That year the WHO spent over $US 13 million on malaria eradication, but only $US 64,000 on smallpox.

The expenditure on smallpox increased with each subsequent year, but even by 1966 it was only $US 426,000, still far too little. Between 1959 and 1966, the WHO spent only about $US 2.3 million on smallpox, for an annual average of $US 300,000, including vaccine contributions, in contrast to the almost $US 2 billion invested in the malaria effort. According to D. A. Henderson, who ultimately led the smallpox eradication program, "The Director-General at the time was quite clever in the way he handled the pressure to go ahead and do the [smallpox] eradication program. He would increase the WHO budget by an amount to reflect current inflation, which was running quite high at the time, and then add in $2.4 million dollars for the smallpox eradication campaign. . . . Taken together, this represented a 15–20% increase in the WHO budget, and that got everyone's attention." Most member countries wouldn't go along with such a substantial increase. They were already contributing heavily for malaria, and the member states were simply reluctant to try to sustain two global eradication campaigns at the same time.

The staffing of the campaign was equally indicative of priorities within the organization. From 1959 to 1963, only two employees at WHO worked full-time on smallpox: a medical officer and his secretary. This changed for the worse in 1963, when that officer left WHO and his duties were added to the duties of another staff member; directing the global smallpox eradication campaign was made, if only temporarily, a part-time job. One critic even suggested that smallpox eradication was a "foster child," while malaria eradication was the "favourite daughter of WHO."

One reason for reluctance on the part of the member nations to support the eradication effort may have been concern for the reputation of WHO itself. According to the official history of smallpox eradication published by WHO in 1988, the early 1960s were a difficult time for the organization. WHO was less than 20 years old and was still making a place for itself in the world. It had invested heavily in malaria eradication, both in terms of money and credibility, yet by 1963 that "programme was already recognized to be in trouble." Setbacks in the fight against malaria—not to mention the prospect of outright failure—were a serious concern for WHO as an institution and made a second eradication campaign look only that much more risky.

Malaria eradication may have made it harder for smallpox eradication to get off the ground in several other respects as well. When WHO decided to convene an Expert Committee on Smallpox in 1964, the course of action outlined by that committee was strongly influenced by the strategies being used against malaria. Just as there had been an attack phase against malaria, so there was to be an attack phase against smallpox. And just as attacking malaria had meant spraying each and every pertinent building with DDT, so the same phase against smallpox meant immunizing 100% of the population in endemic regions.

The opinion of the Expert Committee was also no doubt influenced by evidence coming out of India's smallpox effort. When WHO first accepted the challenge of smallpox eradication in 1959, most people in the public health sector thought that an 80% immunization rate would be sufficient to stop transmission of the virus. Evidence from India, however, seemed to tell a different story—that smallpox could persist in spite of vaccination rates *over* 80%. As it turned out, this evidence was quite misleading because the reported rates were based solely on the number of vaccinations administered. If 100,000 vaccinations were administered, in other words, it was assumed that 100,000 people had been immunized. A closer look revealed that many people were being vaccinated two and three times, that some vaccinations did not take, and that in some cases the vaccine itself was inactive. India's rate, later evidence showed, was considerably lower than 80%, but this information was unavailable to the Expert Committee.

The apparent anomaly—continued transmission in spite of high vaccination rates—together with the all-or-nothing precedent of the malaria strategy, helped to set the bar for smallpox eradication at a daunting height. According to the WHO Expert Committee, "the target must be to cover 100 per cent of the population." No doubt it was hard to take the idea of vaccinating *every* person in Asia and sub-Saharan Africa particularly seriously.

In spite of the reluctance, the Soviet delegates continued to pressure the organization to spend more on smallpox eradication, to set priorities for the campaign, and especially to gather more reliable data—both about the prevalence of smallpox worldwide and about existing vaccination or eradication campaigns. The latter was an especially important and vexing point. Most reports from endemic countries were suspect and tended to vary as widely as did the effectiveness of their vaccination programs. It was often impossible to know how many vaccinations had been administered or even how many cases of smallpox had occurred in a particular region. At the time, WHO officials estimated that only 5% of all cases were reported, but this figure was eventually pushed down to 1%.

By 1965, momentum to intensify the smallpox effort had begun to build, as both individuals and events—in particular several imported outbreaks in Europe—continued to underscore the costs of *not* eradicating the disease. The development in 1964 of a pedal-powered jet injector, a hydraulic device capable of performing 1,000 vaccinations per hour, made the gargantuan task of vaccinating hundreds of millions of people seem more feasible. WHO created a separate Smallpox Eradication Unit charged with overseeing the campaign.

In November of that year, another significant event transpired. President Lyndon Johnson wanted a program to announce in conjunction with International Cooperation Year for the United Nations. The State Department put out a call for proposals, and one in particular caught the attention of the White House. It was a proposal to establish a measles and smallpox vaccination campaign in 18 African countries where both diseases were endemic. The program fit Johnson's needs and thus it was that the U.S. contributed the money, material, and expertise to set up the African program. USAID provided the money and material, and the CDC provided the expertise. D. A. Henderson, the CDC architect behind the idea, was dispatched to Africa to get the program under way.

At the same time, the U.S. delegates to WHA were under some pressure of their own. The State Department was looking for a way to improve U.S.–Soviet relations, and the smallpox program seemed like a perfect way to make some advances. The U.S. delegates consequently found themselves in a curious position—advocating smallpox eradication while under orders not to vote to increase WHO's budget. Money, therefore, remained a major sticking point.

Up to this point, the director-general of WHO, supported by most of the industrialized member-nations who paid most of the bills at WHO, had hoped that voluntary donations would cover the expenses of the campaign. WHO expenditures on it had been, as already noted, embarrassingly small. But donations had never approached what was necessary, and as the projected cost of a full-fledged program rose, it became clear that donations never would be enough. In 1963, it had been thought that $US 10 million in "external support" would suffice; in 1965 the figure was $US 28 million; by 1966, after a detailed examination of strategies and costs, the director-general reported to the assembly that eradication would cost $US 180 million, of which $US 48.5 million would have to come from international assistance. In light of this reality, and after much debate, the World Health Assembly voted in 1966, by the narrowest of margins, to allocate $US 2.4 million for smallpox. The Intensified Eradication Programme was about to begin.

Commitment, Evolution, Success

After almost a decade of half-starts, head shaking, and WHA resolutions, smallpox eradication began in earnest in 1966, just as malaria eradication was failing. This failure, in fact, made many WHO officials, including the director-general, doubtful not only of eradicating smallpox but of eradicating *any* disease. Eradication was a concept less than a century old, and the idea of eradicating a human disease was less than 50 years old. Malaria eradication, the first global effort to put that idea into practice, had not exactly met expectations. Nevertheless, in 1966 the Smallpox Eradication Programme became the *Intensified* Smallpox Eradication Programme.

The intensified program required a full-time director, and the director-general's less-than-sanguine view of the program ultimately dictated his selection of who would head it. He blamed the U.S. delegates for leaving him with another program that he was fairly certain would fail, and he was determined that the failure would rest at their feet. He therefore insisted that the director for the Smallpox Eradication Programme was going to be from the U.S. The choice of a U.S. representative, of course, wasn't popular with the Soviets, who felt they deserved the responsibility since they had been advocating for the program for so many years. With more than a little reluctance, D. A. Henderson accepted his orders to go to Geneva as the new director of the Intensified Smallpox Eradication Programme when he was summoned. As he later described it, "It was either go or be fired!"

It is no doubt fortunate that Henderson wound up in that role. He had already had 15 months of field experience setting up the USAID measles and smallpox program in West Africa. It was an effort that in some ways shaped the course of the global campaign. His experiences gave him the confidence that eradication was achievable, and he promptly began assembling a hand-picked team to join him. Henderson is quick to point out that he was only one of many, but his forceful personality and leadership style were essential in the years that followed. His elite team included many who went on to assume prominent roles in future initiatives against other diseases.

From that year until 1980, when eradication was certified and declared complete, perhaps the most important characteristic of the men and women who became the "smallpox warriors" was their adaptability. Given the constraints on the program—high-level skepticism, dependence on donations, competition from other health concerns—the people in charge of the program understood from the beginning that compromise and flexibility were essential.

The WHO even had to rethink itself. Prior to the intensified campaign, the primary role of the organization had been to provide countries around the world with technical assistance in various health projects. Furthermore, most of those projects affected only one country, or perhaps two or three. Although it had been running the malaria eradication effort for over ten years, the needs of that effort were quite different from those of the impending smallpox campaign. The organization, in other words, was not well prepared to generate and maintain international financial support for a global campaign or to coordinate the material and personnel of a campaign that would be integrated with national health ministries. Yet with the Intensified Programme, the organization found itself facing exactly those challenges.

The strategy at the beginning of the Intensified Programme had been fairly simple, or at least simply stated: mass vaccinations, followed by intensive surveillance. Both these tasks, of course, were immense. When the Intensified Programme began, the smallpox situation had improved significantly from 1959, but the disease was still endemic in at least 31 countries and threatened over a billion people. Even with jet injectors, vaccinating that many people was a giant task. The same was true of setting up surveillance networks. The first requirement of effective surveillance is extensive, reliable data, but in 1967 such data were hard to find. Even the number of cases was grossly underreported. In the mid-sixties, about 50,000 cases of smallpox were officially reported to WHO—out of an endemic population of over one billion. WHO eventually estimated that the actual number of cases annually was about 10 million, including two million deaths.

This two-phase strategy—massive vaccination and intensive surveillance—was subject to revision, however, and this was an important difference between smallpox eradication and its failing predecessor, malaria eradication. Unlike their counterparts in MEP, the leaders of SEP did not stake everything on a single plan. They continued to gather information and to refashion their program in the light of that new information. This willingness to learn through the innovation of the field staff and to transfer its discoveries into the broader campaign allowed the Smallpox Eradication Programme to evolve a program that worked.

The chief revision in strategy came in 1968, one year into the Intensified Programme, with the shift away from mass vaccination and toward "surveillance/containment," or "containment vaccination." Prior to this, smallpox eradication had been synonymous with mass vaccination. Technical advisors had assumed that smallpox was so highly contagious that it would be spread throughout any endemic region—worse in some places, but bad everywhere. Focused strikes were impossible against an enemy so widely dispersed. The best way to proceed was to cover the population with a blanket of immunity—to vaccinate 100% of the population.

In 1968, however, it became clear that the flat-footed strategy of mass vaccination was not the best one. Smallpox was not as highly contagious as had been thought. A person was infectious only during the rash phase, and even then only to face-to-face contacts; typically the person transmitted the disease to no more than two or three others, usually within the person's household. As a result, smallpox outbreaks developed relatively slowly and tended to cluster. Even in countries with high rates of infection, the disease tended to concentrate in only 1% or 2% of the villages.

Because of these characteristics, it was possible to fight smallpox in a more mobile, aggressive way. This was proven first in West and central Africa. There, under the leadership of Bill Foege, smallpox workers switched from passive to active surveillance: they stopped waiting for unreliable reports to trickle in and began to gather information about the disease from as many sources as possible. They solicited information, using newspapers, radio announcements, and letters, talking to missionaries, visiting village elders, and in some areas searching house to house. This form of surveillance uncovered many more cases than had previously been reported, but it also made it possible for the first time to concentrate personnel and vaccine on actual outbreaks. Having defined the area of each outbreak, the smallpox workers then vaccinated everyone within that area. In essence, the surveillance/containment treated each outbreak as if it were a fire and then built a firewall of immunity around it. With no new host to infect, the virus could only burn out.

Essentially the same strategy had been used for decades as a control measure in nonendemic countries, such as the U.S. or Europe, where overall immunity was high. What was new and important about the results from Africa was the proof that surveillance/containment also worked in highly endemic countries where immunity was low. Although some national authorities initially voiced strong resistance to this switch in strategies, acceptance did finally come. Organizers of the program now saw that it was possible to break transmission even though vaccination rates might be as low as 50% or 60%. Mali, for example, with only 51% of its population vaccinated, eliminated smallpox in 1969. Sierra Leone, which had the highest smallpox rate in the world in 1968, eliminated the disease in 1969, with only 66% coverage.

Not only did the strategy evolve, but the tools did as well. Two significant discoveries made the mass vaccination campaigns and surveillance/containment immunization efforts viable—one improved the vaccine, and the other improved the procedure of immunization itself.

The vaccine innovation made it possible to conduct immunization in the remotest villages and in the most primitive of environments. The vaccine being used at the beginning of the intensified program was anything but suitable for fieldwork in such challenging settings. It was supplied as a liquid, which was stable for no more than 48 hours and easily contaminated. Researchers at the Lister Institute in London went to work with very limited resources to find a different approach. The researchers found that freeze-drying the vaccine and storing it in heat-sealed ampoules maintained its potency and stability for long periods of time. The ampoules could be carried in a shirt pocket for weeks and then reconstituted when they were needed. Manufacturing practices were speedily changed, and the improved vaccine was quickly adopted.

The improvement in the procedure used to deliver the vaccine was the second innovation, and it allowed field staff and volunteers with little training to conduct immunizations. The procedure in use at the outset of the intensified campaign relied on scratching the virus into the superficial layers of the skin, a time-consuming process that sometimes resulted in serious wounds. Recognizing that a more efficient method was necessary, the plan was to use the Ped-O-Jet, the pedal-powered jet injector, to speed vaccination. Once the device was in the field, however, it proved unreliable and expensive, not to mention difficult to transport into remote areas.

The solution came from the unlikely collaboration of a microbiologist working for Wyeth Laboratories, Benjamin Rubin, and a sewing machine manufacturer, Gus Chakros. The solution, a bifurcated needle, was nothing more than a sewing needle with the loop ground down into a two-pronged fork. A tiny piece of wire suspended between the two prongs created a device that delivered a very precise amount of vaccine. When a vaccinator made a series of small jabs with the prongs, the needle scratched the skin and delivered the vaccine safely. By 1970, every vaccinator came equipped with the new needles and the freeze-dried vaccine.

Smaller innovations also grew out of the ingenuity of a creative field staff and a collegial organization that not only fostered independent problem solving, but also was prepared to institutionalize the results. Simple things such as smallpox recognition cards and rumor registers were described in the regular surveillance newsletters and in the periodic review meetings. Field training for epidemiologists and a way to evaluate performance and offer timely remediation when standards weren't met were additional hallmarks of the smallpox eradication effort that differentiated it from the now abandoned malaria effort.

This management approach—relying on experimentation and innovation rather than assuming that everything pertinent had been discovered—didn't come without a fight. In 1966, the newly established Smallpox Eradication unit, the office within the WHO charged with organizing the global campaign, found its budget cut drastically, even as the Intensified Programme was getting under way. There was suddenly less money than had been expected for travel, for consultants, for scientific meetings, and for basic research.

In part, these reductions reflected the continuing skepticism of some senior WHO officials, including Director-General Marcolino Candau, about smallpox eradication. The cuts were a kind of foot dragging. Worse than that, they indicated a fundamental misconception, the same one that had weakened the malaria program: the plan for eradication, the reduced budget seemed to say, had already been laid out—everything that the organizers needed to know was known. The money was restored only after Henderson, having just assumed his role as the chief of the Smallpox Eradication Unit, threatened to resign over the cuts. The ultimate success of the program proved that Henderson was right to insist.

Skepticism had its lighter side as well. One WHO official, after viewing what might only be described as the freewheeling approach of the teams operating in India, commented that if the India campaign were successful, he would "eat a tire off a jeep." When the last case from India was reported, Henderson sent him the jeep tire.

Smallpox Zero

Every endemic country could tell its own dramatic story about fighting smallpox. Local and bureaucratic resistance, underfunding, understaffing, false alarms, miscommunication, setbacks—the 44 countries that eventually participated in the Smallpox Eradication Programme faced each of these challenges and many others. None of the battles was easy, and each one illustrated the combination of determination, expertise, sacrifice, and ingenuity that, by the end of the 1970s, had finally stopped the transmission of smallpox.

It was in Ethiopia, however, that the Smallpox Eradication Programme came closest to failure. On the one hand, many of the difficulties encountered there were typical: a substantial population, widely scattered over rugged terrain; a proliferation of languages; a dearth of roads, and most of those poor; few health workers, fewer clinics; relentless rain from June to September; abiding suspicion of outsiders; and variolation, which not only offered a traditional alternative to the outsiders' innovation, but also helped to spread the disease. These were the common challenges smallpox eradication faced throughout many of the remaining endemic countries. On the other hand, Ethiopia also presented obstacles that were more purely political: indifference from the health ministry; active resistance from the malaria eradication bureaucracy; and armed rebels along the frontiers. Tangled together, these problems made eradication in Ethiopia uniquely difficult.

The first problem was getting started at all. The WHO began its Intensified Smallpox Eradication Programme in 1967, and by 1970 a heartening number of countries had been able to break transmission of the disease. To the south and west of Ethiopia, serious progress had been made in Sudan and Kenya, and to the east, Somalia was smallpox free. But in Ethiopia, where many thousands of cases occurred each year, the government declined even to discuss the topic. Repeated offers by the WHO smallpox eradication staff to begin exploring the requirements of an eradication program were refused.

As odd as it may seem today, this refusal was not capricious or perverse. It was based, at least in part, on an assessment of the best interests of a very poor country. For the eradicators at WHO, Ethiopia was an island of smallpox surrounded by countries that were or soon would be smallpox free. From their point of view, Ethiopia threatened all the progress that had been made in the region. But from the point of view of the Ethiopian Health Ministry, smallpox simply wasn't that important a problem. In a poor country, where people struggle every day just to eat, the interests of the region or the globe may not seem very pressing. Besides, the disease, although widespread, was mild; when eradication finally got under way in 1971, variola minor was the only disease discovered. Thus, the one advantage of smallpox in Ethiopia—its low mortality rate—turned out, for the eradication program, to be a liability. The Health Ministry had too many other competing priorities: hunger, infant mortality, cholera, and malaria, to name only a few.

Another competing priority, and perhaps the most important one, was malaria eradication. Although the global malaria eradication campaign was on the verge of failure by the late 1960s, it was still a powerful presence in many countries, not only as a fighter of disease but also as an institution. Like any bureaucracy, it could be protective of its interests and suspicious of rivals, real or imagined. In Ethiopia as elsewhere, malaria eradication was separate from the rest of the national health service, and with over 8,000 people on staff, it was also far larger. Its annual budget, most of which came from donor nations, was $US 8 million, and this money, like the army of workers, was devoted entirely to malaria.

Given the importance of malaria, it's easy to understand and even applaud this degree of institutional focus. It's less easy to understand why smallpox eradication was seen as a rival, although it clearly was viewed as one. Smallpox eradication, unlike malaria eradication, was relatively simple and inexpensive, and it normally integrated fairly easily into an existing health system. Although the WHO smallpox staff might ask for help from other health services—with the reporting of suspected cases, for example, or with educating the public about vaccination—they gave help as well, particularly when other vaccination campaigns were taking place. The resources for smallpox eradication did not have to come from the malaria eradication budget.

Nevertheless, the malaria staff in Ethiopia believed that there was room for only one eradication program and had convinced the minister of health to reject the overtures from the Smallpox Eradication Unit. They argued that every dollar spent on smallpox was a dollar not spent on malaria. The Ethiopian government, having watched malaria eradication drag on and on, was understandably doubtful about a similar program anyway, even though the government was responsible for only a fraction of the expense.

Things remained at this impasse for three years, until the smallpox staff at WHO headquarters found two unexpected paths around the minister of health. The first appeared when the minister was out of the country. The WHO had been periodically renewing its request for permission for the SEP team to make an exploratory visit. A deputy minister had apparently failed to get the message that smallpox was of no interest, and in the minister's absence, the deputy said yes to WHO. Amazed, Henderson and another member of the smallpox staff rushed to Ethiopia. Henderson remembers the event with clarity: "I arrived the day after the Minister arrived. He was very unhappy about seeing me, and that's when we sort of worked through what I could offer as a carrot. Well, I could offer some vehicles, but they already had all these UNICEF vehicles, parked out in great rows. You could see them from the office. We could offer some additional funding to pay for staff time, but it wasn't a lot of money." They had so much money and equipment in the malaria program, in fact, that the carrots simply weren't appealing.

Nevertheless, the minister of health allowed the team two weeks to prepare a plan. The second break came after the plan was finished. The minister, predictably, was unimpressed, but Henderson had also given a copy to a friend and colleague, Kurt Weithaler, who was in Ethiopia serving as the director of the hospital for the Emperor's Imperial Guard. Weithaler in turn gave the plan directly to *his* friend, Emperor Haile Selassie. The emperor approved the plan.

Gaining approval of the plan was, however, only the first of many challenges. In 1970, Ethiopia was treacherous terrain. The country comprised slightly more than 470,000 square miles—not quite as large as Texas, Oklahoma, Kansas, and Nebraska combined. Twenty-five million people lived there, and in the south nomads wandered freely into Somalia and Sudan and back again. As was often the case in a developing country, it was hard for someone at one end of Ethiopia to communicate with someone at the other end. Not only were the phones unreliable and the postal service, too, but huge expanses of the land had no roads at all. Even bus travel was impossible. The smallpox staff members often found themselves walking many hours to investigate a single suspected case, or, on a good day, riding a mule.

Some health services were available, but these were concentrated in the capital, Addis Ababa, and in Eritrea, where first the Italians and then the English had invested modestly in clinics and hospitals. Five percent of the population had access to these services. The rest made do without. For most Ethiopians, even the closest clinic was a full day's journey away; 40% lived more than three days away from any formal health care.

In other words, Ethiopia had little upon which the smallpox eradication staff could build. And to make the job even harder, in January of 1971, when the program began, there were only 39 staff members to do the building. Across town, malaria eradication had 8,000 people on staff. Its fieldworkers made routine house-to-house visits and were thus perfectly positioned to help with surveillance, but this assistance was refused. The malaria program also had fleets of unused vehicles, but when the smallpox staff asked to borrow a few—it had only six of its own and was responsible for almost half a million square miles—the vehicles mysteriously disappeared.

One of the first tasks in building the program was to draw an accurate picture of the smallpox problem. How prevalent was it? How severe? Prior to 1971, data about the disease were, like the health system itself, incomplete and unreliable. A few hundred cases were reported each year—197 in 1969, for example—but because these were only the cases seen at hospitals, there was good reason to believe that the actual number might be higher.

This was, in fact, the case. In the last two weeks of January, 278 cases were discovered; 1,493 cases in February; and 3,434 in March. By June, Ethiopia had 13,447 documented cases, more than the total number of cases reported the previous year in India, a country of over 500 million people. By the end of the year, the total number was 26,329 cases.

Because these were cases of variola minor, infected people were usually not so sick that they had to stay in bed. This, of course, meant that there were relatively few deaths (530 in 1971), but it also meant that most people could work and travel with the disease, and thus spread it. In March, smallpox imported from Ethiopia caused an outbreak in Kenya, underscoring both the danger Ethiopia posed to its neighbors and the difficulty awaiting the smallpox staff.

However daunting the problem may have appeared, the astonishing number of cases proved that the staff, strapped though it was for resources and personnel, was making progress. The surveillance system, now overseen by Ciro deQuadros from Brazil's successful eradication effort, was obviously working, and it was thus possible to target the work more effectively. The field staff administered slightly more than three million vaccinations during that first year, primarily in the four southwestern provinces where the health service structure was slightly better and the acceptance of the vaccine was reasonably good.

During the first half of 1972, the number of reported cases remained high, but in the second half the number began to fall. By the end of the year, fewer than 1,000 cases were being reported each month. This decrease was especially significant because by the end of 1972 the surveillance system was larger and more experienced than it had been the year before. Yet the harder the surveillance teams looked, the less they found; the total for the year was 16,999 cases—still high, but far lower than the year before.

More good news followed. Surveillance teams reported that transmission had been interrupted in two of the four targeted southwestern provinces—they could find no cases there. The most surprising news of all, however, came from the two northernmost provinces, Eritrea and Tigray. Because of staffing shortages, not to mention civil war, very little change had been expected in this region. Nevertheless, with only one sanitarian per province, Eritrea and Tigray saw their last endemic cases in 1972.

Things had gone so well, in fact, that it didn't seem unreasonable to hope that 1973 might be the last year of the campaign. The work rolled on.

Teams continued to cover vast tracts of lands, often on foot, visiting markets, clinics, churches, and schools, investigating cases and vaccinating those who were willing. Many, however, were not. People were often quite hard to reach, literally and figuratively, especially in the rugged central provinces. In some cases no amount of persuasion, from the team or from priests and village elders, could convince everyone that vaccination was a good idea. A team might spend weeks in a single village, yet leave with only half the people vaccinated. Even so, the number of cases reported each month continued to fall. By September the number was down to 71.

The fall of that year, however, brought a major setback. A drought led to a famine in several provinces, which in turn created a human catastrophe: 200,000 deaths. Compared with the famine, of course, every other problem in Ethiopia at that time seems trivial, or secondary at best. Still, there *were* other problems, and for the smallpox staff the problem was refugees. As people fled the famine-stricken regions, they took smallpox with them into other provinces and even into Somalia. The number of cases discovered in November and December jumped by 700% from what it had been in September.

Complexities continued to multiply. In 1974, civil unrest grew into revolution and Haile Selassie was overthrown; as a result, certain areas were so unsafe that eradication teams could travel only with security escorts, and sometimes not at all. Guerrilla fighting broke out in the east between Ethiopia and Somalia. A key WHO program adviser, Dr. Petrus Aswin Kosara, died of a heart attack. He was only 43 and had just returned after weeks of strenuous work planning a new helicopter-assisted search program. His death took an emotional as well as physical toll on the team. And both money and manpower, more necessary than ever as the work became more difficult, were still scarce. International donors and volunteer organizations feared the political turmoil in the country, and WHO was sending every spare dollar to Bangladesh and India in an attempt to wipe out the last cases of variola major.

In spite of all this, however, the number of cases continued to fall, and the work intensified. Teams from Kenya, the Sudan, and French Somaliland searched some of the most difficult regions in the east and the south. Meanwhile, the eradication staff put every available volunteer to work, including relief workers sent to treat victims of the famine. Even the malaria staff, after years of remaining aloof, began to pitch in. In November, the arrival of two helicopters opened up even the most inaccessible highland regions. Teams were set down and then, after ten days of trekking, investigating, and vaccinating, picked up again. Known as

Operation Crocodile, this aerial part of the program not only moved workers efficiently, it also was so dramatic a spectacle that it tended to break down resistance to vaccination. Anyone who could fly through the air like that must know what he's talking about.

In 1975, India and Bangladesh stopped transmission and recorded the world's last cases of variola major. This was good news for the world—*major* was the severe form of the disease—but especially good news for Ethiopian eradication. Now that Ethiopia was the last country on earth with smallpox, more resources became available to fight it there. The Soviet Union donated vaccine. Money from Western countries and the WHO made it possible to expand the staff and secure two more helicopters. As the staff grew it became harder to supervise and coordinate, but any disadvantages were offset by the ability to focus more intensely on the remaining endemic regions.

Challenges continued. Some international volunteers left the country because many of the regions continued to be dangerous; in the west-central highlands of Gojam, in fact, two Ethiopian vaccinators were abducted and killed by dissidents. One of the helicopter pilots was taken hostage. Still, the revolutionary government decreed that eradication was of the highest priority, and more and more Ethiopian volunteers joined the work.

As more provinces became smallpox free, attention shifted southeast, to the Ogaden desert, where the nomadic ways of the population, not to mention the guerrilla actions of the Western Somalia Liberation Front, made work extremely difficult. It was then, in 1976, that the staff learned that the situation in the Ogaden was much worse than they had been led to believe. On both the Ethiopian and the Somali side of the desert, leadership had been, as the WHO history puts it, "seriously deficient"; fewer people had been vaccinated than had been claimed, and surveillance and containment were often only nominal. For example, an isolation hut—designed, as its name suggests, to keep smallpox victims strictly separated from everyone else—might be in use but unattended, so that the sick and the well, the vaccinated and the unvaccinated, could come and go as they pleased.

In the Ogaden desert smallpox made its last stand in Ethiopia. In spite of a severe flood in the region and brutal fighting between Ethiopian and guerrilla troops, teams were finally able to enter the southern half of the Bale province in July 1976. In the village of Dimo they discovered an outbreak, 16 cases in all, the last of which was a three-year-old girl, Amina Salat.

As week after week passed and it became clear that Amina Salat was the last case in Ethiopia, the eradication program staff held its collective breath. Ethiopia was the last endemic country. Didn't that mean that the last case in Ethiopia was the last one on earth? No one wanted to announce victory prematurely, but after so many years and so much struggle, it must have been hard to resist the urge to shout. A press conference was scheduled for the end of October.

As it turned out, however, Ethiopia was not the last endemic place. In September, Henderson and his colleagues received the news of smallpox in Mogadishu, the capital of Somalia. Despite their disappointment, they had good reason to be surprised as well: for 14 years, the only cases in Somalia had been imported ones, but investigation revealed that these new cases were endemic. Because of the porous Ogaden desert and the sloppy eradication work done there, the disease had slipped back into Somalia and reestablished itself.

The eradication staff found itself not at a press conference announcing the end of smallpox, but back at work—searching, isolating, vaccinating. The work was particularly difficult at the beginning of the crisis because the Somali staff would not admit what it already knew: that smallpox was endemic again in Somalia. Instead, they blamed the new cases in Mogadishu on Ethiopia. This was plausible, and a good deal of time and money were spent searching the Ethiopian side of the Ogaden desert, looking for the source of these cases. None was ever found; they had, in fact, originated in Somalia. The Somali staff knew the truth but, understandably embarrassed, they hoped that no one else would have to know it. This secretive and misleading approach to the problem resulted in several epidemics, and in May 1977 the government was forced to declare an emergency. Between September 1976, when the first endemic cases were reported in Somalia, and November 1977, when Ali Maow Maalin recovered, over 3,000 cases were recorded. Thirteen of those died.

Maalin's was not the last case of smallpox—in 1978 there were two more, one of them fatal—but his was the last *naturally occurring* case. That was the history-making achievement of the Smallpox Eradication Programme: it removed the virus from nature—it stopped transmission for every human being on the planet. For ten years the program had pursued the virus the way hunters pursue a fox, relentlessly, from hiding place to hiding place, until the creature has nowhere left to run. In Ali Maow Maalin, the eradication program finally brought smallpox to bay. When Maalin was placed in isolation, there was no one new to infect—no way of escape. And when he recovered, that was it. There was no more wild smallpox.

The Smallpox Dividend

Locking the lid on smallpox has had enormous consequences for the entire globe. The most obvious benefit derived from this achievement is the reduction in human suffering and loss of life for people in all countries. When the intensified eradication effort began, the WHO estimated that ten million people were afflicted with smallpox every year and two million died from the disease. Almost all these people lived in countries without the resources to provide the protection that the more advantaged had. The SEP leveled the playing field. In the more than two decades since the last case of smallpox, 240 million people have been spared this disease, 48 million of whom would likely have died. It is impossible to quantify the value of this benefit in any meaningful way.

In dollars, the savings are staggering. By 1972, the U.S. ceased vaccinating its children against smallpox. Other nonendemic countries followed suit. When the world was certified free of the disease in 1980, the WHO issued its recommendation that all vaccination be discontinued throughout the world. By 1984, no one was administering smallpox vaccinations. With disappearance of the disease, quarantine measures were no longer necessary. Current estimates place global savings at $US 1 billion—annually.

The total cost of the ten-year eradication program for consistency was approximately $US 312 million, which seems a small investment compared with the enormous benefit. Very little of this money came from cash donations to the WHO. In fact, as late as 1973, the total cash contributed amounted to only $US 79,500. Several countries did donate vaccine, but the bulk of the monies came from the coffers of national governments or through international assistance.

It is difficult, even impossible, to know whether those savings have actually been shifted to other public health programs. The need associated with advancing global health through disease control remains vast. Malaria, tuberculosis, and HIV represent serious challenges in many parts of the world. Diarrheal and respiratory diseases plague the world's children. All act to hamper economic development in both direct and more subtle ways. Whether the resources saved have been transferred to health or non-health uses, they are at least resources freed and available for other purposes.

While the impact of financial savings derived from smallpox eradication may be difficult to track, the contribution that the eradication campaign made toward building a public health infrastructure is more tangible. Because the program had to be run with such relatively limited resources, there was no choice but to integrate it into each country's existing health services structure. There simply wasn't enough money to set up entirely separate services similar to the ones created during the malaria campaign, even if the planners had thought it a good idea to do so. In many of the developing countries, this meant that operational management systems extending into every village were developed and that quality control measures for vaccine stability and potency were introduced. Success also depended heavily upon developing a functioning surveillance system, which was virtually nonexistent in most countries at that time. The smallpox program began to build the infrastructure.

This is in no small part because of the leadership of the program. From the outset, Henderson and others saw the program in a broader context. Vaccines were highly effective interventions. Henderson reflects, "[I]f you could devise a scheme that delivers a vaccine throughout a country, with appropriate quality control and management oversight, it would be like bringing in a Christmas tree. You could hang additional vaccines on it; you might even integrate things like micronutrients into the structure. As we went around and we visited other countries, I was amazed to find whole wards full of children that had measles, polio, etc. and their governments were doing nothing, including [providing] DPT."

Although the desire to expand the smallpox program went unfulfilled during the campaign itself, the groundwork laid by the staff made later efforts easier. The Expanded Program for Immunization, or EPI, as it was known, built upon the accomplishments of the smallpox campaign. EPI's mission was to deliver vaccines against at least six diseases—measles, polio, diphtheria, pertussis, tetanus, and tuberculosis—to children in over 100 countries around the world. This program has grown and still functions today.

In the final analysis, people, not programs, may well be the greatest contribution that the SEP made beyond stopping a deadly disease. Health ministers and their staffs were greeted with new regard by their governments; they were, in fact, getting rid of smallpox in their countries. Henderson is understandably proud of the additional achievement. "We employed a lot of younger people and many of them came out of SEP with a can-do attitude. It was apparent that we could vaccinate large numbers of people. With a little bit of organization you can get 90% of a population. We set a minimum goal in Africa of 500 vaccinations per vaccinator per day and we were reaching this with no trouble. It was just felt to be impossible to do this before we got started, and so there was a feeling that, yeah, you could do this—you could vaccinate a lot of people."

Many of these people have gone on to prominent roles within the public health community: William Foege, Donald Hopkins, Ciro deQuadros, and many others. Some of their voices can be heard at the end of this book.

Sadly, there is also a dark side to the dividend. Today the smallpox virus still exists, as virulent as ever. It can be found in two official repositories, locked away under heavy security. The virus is still dangerous—the two cases in 1978 originated in a laboratory in Birmingham, England. At the Centers for Disease Control and Prevention in Atlanta—one of the two sanctioned repositories—the virus is literally shut up in a freezer draped with chains and locks. The other repository is perhaps less well guarded. It exists at Vector, a huge, financially troubled former bioweapons laboratory in Novosibirsk, Siberia. Suspicions abound that other, nonsanctioned stocks exist. The WHO had planned for destruction of the last virus stocks in 2000. So far, world leaders, including those from the U.S., have blocked its scheduled death.

Heightened public concerns about the intentional release of the smallpox virus have left governments scrambling to develop a response in the event of its use in a bioterrorist attack. The U.S. will spend $500 million to stockpile vaccine and unknown additional monies to prepare its physicians and public health infrastructure against such an event. Other countries in the industrialized world are following suit.

The threat of the virus's release has left many wondering whether eradicating smallpox (or any other disease) was the right thing to do. With the cessation of vaccination two decades ago, practically everyone on the planet now has some susceptibility to the disease. The threat that someone will intentionally release the virus is real, and the threat has taken on an escalating importance in the face of events in the United States that began on September 11, 2001.

The people who have worked long and hard within the public health community to bring about the eradication of smallpox, and those who hope to follow it with the eradication of other infectious diseases, have decided views in terms of the ultimate value of eradication. Walter Dowdle, a former deputy director of the CDC, expresses the opinion of most. "Last week, I was involved in the smallpox training here at CDC. We were training all the responders for possible bioterrorism. You know, the whole question comes up, what does this have to do with eradication, or vice versa, and I think that really has to be put into perspective. My personal view is that this does not in any way influence the overall good that comes from eradication. One just has to balance that. The argument that eradication just gives terrorists another tool, I don't think that is a relevant argument.

"We talked about this for a long time—the need for preparedness, the need for having vaccine stockpiles. I think the problem with smallpox is that we are just scrambling to do things we should have done a long time ago.

"Hopefully, one of these days, we will be on the verge of eradicating other infectious diseases, either by design or by improved global health. What then are the options? Do we actively prevent a disease from being eradicated because its agent might be used by a bioterrorist? Do we insure that the disease continues in poverty stricken countries so that rich countries are not threatened by bioterrorism? That's not exactly a great public health goal. There is no alternative. You can't say I'm not going to get rid of a disease simply because of bioterrorism. It's ridiculous; that option is simply not viable."

An Incomplete Life: the Biologic Perspective

The Smallest and Simplest of Things

Before the recent discovery of the prion—the mysterious protein associated with mad cow disease—viruses were the smallest and simplest infectious agents known. They are so simple, in fact, that the question, "Is a virus alive?" doesn't have a simple answer. Most people, when they think about disease, think about "germs," and they lump viruses and bacteria together in that category. But a virus is actually less than a germ. Compared with a virus, even the single-cell bacterium, the smallest *living* creature, is vast and complex, made up of hundreds of enzymes and structural materials and able to reproduce quite efficiently all on its own. As long as it gets food and water, a bacterium can thrive on your skin or on the sponge in your kitchen sink. More complex still are multicellular parasites, such as the *Plasmodium* species that causes malaria.

A virus, in contrast, is not even a cell; it's a particle of genetic material covered by a protective protein. On the one hand, a virus does not breathe or feed or excrete or exhibit, on its own, any of the attributes of life. As a textbook might put it, a virus is an "obligate intracellular parasite," that is, it depends wholly on a host cell. On the other hand, once inside a host cell—and this can be bad news for the host—a virus begins to display certain characteristics of life. With the help of enzymes stolen from the cell, a virus can replicate itself quite efficiently. It can also mutate and recombine, and so evade the immune system.

A computer virus makes a useful analogy. On its own, outside a computer, it isn't much: just some code on a floppy disk. Inside a computer, however, the virus takes over. If the message of the code is, "Make copies of me until you crash," then the computer makes copies until it crashes.

A biological virus works in a similar way. It's a particle of code, and its only function is to see itself replicated. To do that, it commandeers a host cell. Imagine that a healthy cell is a factory, living and growing and eventually dividing into two healthy cells, then into four, and so on. When a virus invades, it takes over the factory. Some of the machinery the cell would have used to reproduce itself is put instead to the task of making virus copies. The normal, healthy functions of the cell are interrupted, which explains why an infected cell weakens or even dies. Although some viruses are fairly benign—warts and cold sores, for example, are both caused by viruses—some are able to do so much damage to so many important cells that serious, often fatal disease is the result: rabies, AIDS, polio, Ebola, and of course, smallpox.

The Smallpox Virus

The brick-shaped variola virus, which causes smallpox, is unique for at least two reasons. First, variola is one of the largest viruses, each particle measuring around 0.2 by 0.4 micrometers, or about one-thousandth the size of the period at the end of this sentence. Because of its bulk, smallpox was the first virus to be seen (just barely) under an ordinary light microscope. Second, smallpox is among the most complex of the viruses, approaching a primitive bacterial cell in both appearance and biochemistry. At least 100 proteins reside in each particle, including many of the enzymes the virus needs in order to reproduce. Consequently, the smallpox virus carries more of its own vital supplies and relies less on its host cell than do most other viruses.

Actually, two related strains of the virus cause smallpox. Each one produces a distinct form of the disease. The primary distinction is severity. At the mild end of the scale is variola minor, with a fatality rate of 1%. At the other end is variola major, with a fatality rate of 25%–30% and a tendency to scar, sometimes even to blind, its victims. For centuries, variola major was the only type of smallpox known, but for unknown reasons, it began to give way at the end of the 19th century to variola minor.

A smallpox virus, traveling through the air in a microscopic water droplet, is covered by an outer surface that contains some materials (lipids and proteins) co-opted from its previous host cell. When inhaled, the virus immediately attaches to one of the cells that line the throat, fusing its outer layer with that of the new cell and crossing quickly inside. Once inside, the virus sheds its outside coating and sets to work. Within hours the infected cell has begun to synthesize the component proteins of the virus, and these components then assemble themselves into new particles. The process repeats, over and over, producing as many as 10,000 copies of a single invader before the host cell dies and releases the newly minted and fully infectious particles.

As with other viruses, many of the proteins in the smallpox virus's outer layer will eventually be recognized by our immune system. If a victim doesn't die, his immune system is prepared to fight off any future attacks by the virus.

On the Eve of Destruction

Once smallpox had been eradicated, international officials faced an unprecedented question. What should be done with the remaining stocks of the virus? More than two decades later, there is still no final answer to that question. There are only two official repositories, one in the U.S. and the other in Russia. The CDC in Atlanta, Georgia, holds 451 strains; the Russians have an additional 50 not present in the U.S. collection. The strains include viruses that cause both variola major and variola minor. Many country representatives and scientists have long advocated destroying the remaining stocks on the grounds that doing so lowers the risk of intentional or unintentional release of the deadly virus. Others have argued as vigorously that the virus must be preserved because there is much we can learn from studying it more thoroughly. In the end, the issue has more to do with global politics and international distrust than it does with science.

The WHO has repeatedly scheduled a date for viral destruction, first in 1993, and then in 1996, 1999, and finally in 2002, but each time the execution order has been delayed. The biggest obstacles have been concern about the use of the virus as a biological weapon and concern over the poor state of global preparedness in the event that such a weapon is used. Although only the U.S. and Russia are sanctioned to maintain the virus, it is naïve to think that these two nations possess the only stocks. Too much evidence points to the contrary.

In the years since eradication, researchers in the U.S. and Russia have addressed many of the key scientific questions growing out of these concerns: several antivirals that are likely to be effective have been identified, methods for vaccine production have been improved, vaccine has been stockpiled, and better diagnostic tools have been developed. Multiple strains of the virus have been fully sequenced and cloned, and a reasonable understanding of the virus's virulence factors has been developed.

These advances have left the world better prepared should the virus be used as a weapon, but support for viral destruction is still not forthcoming, particularly from the U.S. and Russia. The current argument for preserving the virus is to continue research: there may be more effective antiviral drugs, a better vaccine, and an animal model that can simulate human disease. On the surface, such arguments seem to make sense. There is, however, a major sticking point—the smallpox virus is strictly a human pathogen, and humans, because of eradication, no longer get the disease. That means that any new vaccine or antiviral cannot be tested in infected humans; hence the effectiveness of new vaccines or antivirals can never be anything more than inconclusive. Despite recent efforts, there is no animal model that mimics human disease. And such research, whether it's the search for vaccines and drugs or the quest for an animal model, comes with a high price tag.

If suspicions about the virus are true—that smallpox is in the hands of a few rogue nations or just plain rogues—then destruction of the virus stocks in the U.S. and Russia are merely symbolic. Yet there is power in symbolism; destroying the stocks would send a strong signal that civilized society views the retention and use of the virus as morally repugnant. Further research efforts to develop antivirals or vaccines that can't be tested in human smallpox are perhaps symbols themselves, for we have a proven vaccine and we have a proven strategy for containment, a strategy that worked so well that the disease was eradicated.

To continue to delay the destruction of the virus at this point seems hard to justify, even if it is a symbolic gesture. The more we work with the live virus, the greater the risk that it will be accidentally reintroduced or fall into the hands of less than scrupulous people. Given that resources are limited and that the real health needs of the world press us on every side, it seems even harder to justify pouring valuable public health resources into additional research against a disease that hasn't occurred for more than 20 years, on the remote chance that it will be used as a biological weapon.

Polio: the Rise and Fall of a Disease

We can see polio more clearly than most diseases because its rise and fall took place within a single lifetime.

John Rowan Wilson, *Margin of Safety* (1963)

When we began the eradication effort in 1988, polio paralyzed more than 1,000 children a day. In 2001, there were far fewer than 1,000 cases for the entire year. But we're not finished yet, and the past year has reminded us that we live in a world where security and access to children cannot be guaranteed. So I urge the world to finish the job. Eradicate polio while we still have the opportunity.

Gro Harlem Brundtland,
Director-General of the World Health Organization (2002)

It is Wednesday, September 27, 2000. In Conference Room 2 at the United Nations, Thaddeus sits quietly in his wheelchair, his face impassive. He has been carefully positioned in front of an enormous box—12 feet high, with foot-high red numbers and an Omega logo. The logo suggests that it's a clock, but it's clearly not one that measures minutes or hours. The readout at the top says 166,024,816.

Thaddeus is ten and a native of India. Before coming to the United States, he lived in an orphanage, taken there after he was discovered abandoned at a railway station somewhere in the city of Calcutta. Thaddeus had no wheelchair then; he moved himself around by scooting along the floor using only his arms. His legs were paralyzed—as they are now—for Thaddeus is the victim of polio.

When Thaddeus was five years old, Mia Farrow, an American actor with an especially big heart, made him a part of her family. She had already adopted many children with special needs, but for Thaddeus she felt an immediate connection. She had lived through the same disease.

Mia Farrow and Thaddeus are here as Special Representatives for the United Nations Children's Fund—UNICEF. The occasion is the Global Polio Partners Summit. Today Thaddeus and his adoptive mother join leaders of government, business, public health agencies, and philanthropic organizations from around the world to start the giant clock that will count down over five years' worth of seconds—the foot-high numbers reading 166,024,816—until polio joins smallpox in the history books as a thing of the past.

The room is filled with influential people—Kofi Annan, secretary general of the United Nations; Gro Brundtland, director-general for the World Health Organization; Carol Bellamy, executive director for UNICEF; Donna Shalala, U.S. Secretary of Health and Human Services, and many others. Ted Turner is here, and so is Frank Devlyn, president of Rotary International, the organization that has made the eradication of polio its primary philanthropic goal.

At 10:00 A.M. sharp, with little fanfare and to little applause, Thaddeus Farrow and Kofi Annan start the Countdown Clock. Then there are speeches. Annan begins by asking the attendees to imagine the last child in the world to have polio: most likely an African child, almost certainly poor, in a country at war—Sierra Leone, perhaps, or Sudan or Angola. Eradicating polio, he says, will depend on reaching this child, and reaching him will be not only hard but dangerous work. He talks of temporary ceasefires and national immunization days. With the numbers counting down behind him he says, "Our race to reach the last child is a race against time. If we do not seize the chance now, the virus will regain its grip and the opportunity will elude us forever."

Next, the director-general of WHO, Gro Brundtland, recounts the progress made so far. By 1988, when the goal of polio eradication was first announced, the disease had already been eliminated from the Americas and the Western Pacific. Still, there were at least 350,000 cases reported each year from the remainder of the world. "Last year, it was 7,000," she says, "and so far this year, less than 2,000. The hardest part has yet to come. We know what we have to do. We have the tools and the strategy. The challenges can be surmounted, but only if current and new partners commit their support through 2005. I urge you all to play your part in making history."

One by one, representatives stand to pledge their support—Rotarians who have volunteered throughout the world; representatives from the health ministries of some 20 countries where polio still exists; and representatives for Japan, USAID, Voice of America, the International Red Cross, and Red Crescent. There is a French pharmaceutical manufacturer who provides vaccine. Donna Shalala promises the continuing support of the U.S. government. "No nation is truly free from polio until every nation is free from polio," she says. Carol Bellamy speaks on behalf of UNICEF: "We stand at the most opportune moment imaginable for attaining the goals for child survival and development set at the 1990 World Summit for Children. Let the eradication of polio herald that reality."

Other notables appear via videotape. Nigerian President Olusegun Obasanjo commits his country—more than 100 million people—to a regionwide surge in eradication activities that will cover 17 countries in west and central Africa the following month; Bill Gates, cofounder of the Bill and Melinda Gates Foundation, which has provided millions of dollars toward childhood immunization, gives polio vaccine to children in New Delhi; Martina Hingis, a WHO Spokesperson for Polio, smashes serves across the tennis net and shares the screen with images of disabled children; and Sheikh Hasina, prime minister of Bangladesh, celebrates the fact that her country has no naturally occurring cases of polio today and wonders, "Who would have believed this could have happened 15 years ago?"

The most affecting words, however, come from Mia Farrow, who talks not about Thaddeus but about herself and the things that happened to her when she was nine years old. "I just fell to the ground one Sunday morning, and I couldn't get up." She spent a month in the hospital, surrounded by other children afflicted with polio—some better off, some worse. She remembers the whooshing of the iron lung and the terror of being placed inside one herself, unable to move, even to scratch her nose. "You lie there. There are holes in the side of it, so that the nurses can reach in when they need to, and I remember screaming for someone to help me." And the ordeal didn't end when she left the hospital. "I knew I had done something terrible to my family. People looked at me with such terror. My siblings were moved out of the house. My house was repainted. The swimming pool was drained. The lawn was torn up and reseeded. The dog was given away. Everyone was afraid of me."

It's hard to imagine today that polio victims believed they had to worry about cleaning the pool or reseeding the lawn, but it's easy to believe that Farrow's helplessness, fear, bewilderment, and guilt were nearly universal. She testifies to the human cost of the disease and then rolls all the speeches and appeals into a single sentence: "Hopefully, soon no one will know about polio."

As the summit draws to an end, the attendees stand to read together the Polio Pledge:

> The world has the knowledge, resources and capacity to eradicate polio by 2005, ending forever the threat of disability this disease poses to millions of people. Eradication is within our reach, and we who are gathered here at the United Nation . . . fully commit ourselves to this moral and historic imperative. We pledge to overcome all remaining obstacles and achieve a polio-free world.
>
> *With one voice, we urge friends and partners everywhere . . .*

In spite of the unity and the pledges of support, however, and the optimistic talk about victory over disease, this summit has a sober mood, and not just because polio is, after all, a sober subject. Although no one confronts it directly today, one fact weighs on everyone's mind: The eradication of polio, in spite of the genuine progress made so far, is by no means certain. The foot-high red numbers might wind down to zero and that last child might still be years away, in spite of every effort. In fact, the clock itself is a reminder that things are not going quite as well as had been hoped. The polio eradication campaign began in 1988; the original goal for its completion was set for 2003. Today the clock is not being started so much as reset and restarted. Officials have had to give themselves two extra years. Like the malaria campaign, which was supposed to last seven years and was called off after 14, polio eradication has begun to stretch out.

This isn't exactly bad news, but it isn't good, either. As an eradication campaign grows longer, time can become an outright enemy, as stubborn as the disease itself. People get tired; they begin to lose a sense of purpose. Governments turn to other problems because there are always other problems. Money gets more difficult to find. When Kofi Annan drew attention to the time slipping away and warned that the eradication of polio might "elude us forever," he was pointing to a political, not a biological, danger. The virus will likely stay the same, and the vaccine that prevents it, too, but the political will to fight the disease is not everlasting.

Like malaria and smallpox eradication, polio eradication is both a medical and a political enterprise. It's concerned with a specific pathogen and a specific set of victims, but it's also a test of international resolve—a test of the very notion of global eradication. This is a double burden. The failure of malaria eradication, leaving aside what it meant for the victims of malaria, made smallpox eradication less likely politically. Governments and agencies had paid a high price, and though they had gotten significant reductions in the malaria burden, they had not gotten what they were promised: eradication. Having lost one war, most weren't eager to declare another.

On the other hand, the success of smallpox eradication smoothed the way for polio. Today, polio eradication finds itself between those poles, making amazing progress yet feeling the strain of a long campaign. It isn't yet clear what will be left of polio, or polio eradication, or eradication itself, 166,024,816 seconds from now. What is certain, however, is that if members of the WHO ever argue the merits of another eradication campaign—against measles, for example—they will be thinking about polio when they do.

Biologic Realities

At the time that the United Nations Polio Conference was taking place, another series of events were unfolding. On January 8, 2001, the Pan American Health Organization (PAHO) sent out a press release announcing the first outbreak of polio in the Americas since 1991. The epidemic was taking place on the Caribbean island of Hispaniola, an island divided between two nations—Haiti and the Dominican Republic. It confronted public health authorities around the world with one of their worst fears, heralding a new difficulty in achieving global eradication of the disease : the outbreak was caused by a variant of the vaccine strain of the poliovirus used to immunize the world's children.

Twenty-one children had succumbed to the virus by September 18, 2001. The youngest among them was nine months old, the oldest 14. Two died; the rest were left paralyzed. An isolated case of polio caused by the vaccine virus isn't new, but the vaccine-derived poliovirus circulating in Hispaniola was unlike any discovered to date.

The polio vaccine contains live virus, attenuated in its ability to cause disease but able to infect and immunize its recipients. In one out of every two million or so children immunized, the vaccine virus reverts to full virulence and causes the disease it's supposed to prevent. What was new about the Hispaniola event was the vaccine strain's recovered capacity to cause serious disease and to spread silently and efficiently from person to person. These new characteristics were gained through both a genetic mutation and the vaccine strain's exchange of genetic information with other related viruses, restoring it to full virulence. The virus had been circulating on the island undetected for perhaps as long as two years by the time it was recognized. Alarmingly, it was clinically indistinguishable from the wild-type virus.

A second alarming fact became evident as investigators from PAHO and the Centers for Disease Control and Prevention tracked the new variant. The percentage of children not adequately immunized against polio in both countries had risen to dangerous levels. Among the 21 affected children, all but one had not been vaccinated at all or had been only partially vaccinated, and there were many others like them. The combination of the revertant and recombined virus with the accumulation of susceptible children was ultimately sufficient to create the conditions for the epidemic.

The health ministries in both countries mounted massive campaigns, going house to house to deliver doses of the oral vaccine to the island's 1.2 million children under five years old. Such campaigns hadn't been conducted for over ten years in the Western hemisphere, and there were logistical problems made worse by heavy rains. After five massive rounds of vaccination—three in the Dominican Republic and two in Haiti—only 40% of the children at greatest risk had been reached. The campaign expanded, now targeting the 2.3 million children on the island who were younger than ten. By the end, 10.8 million doses of vaccine were distributed, and the global community of public health specialists waited nervously for news that the outbreak had been controlled. A warning was sent out to all visitors to the island—"If not vaccinated adequately, you are at risk for polio."

Across the ocean, in March 2001, another outbreak of polio took place. This one was smaller—only two children with the disease had been spotted—first a 13-month-old girl in the Bulgarian seaport of Burgas and then another two-year-old. The WHO medical officer for polio eradication for the European Region, George Oblapenko, knew from the second case that the disease was spreading rapidly and something had to be done. For every child with disease, as many as 200 others were infected with the poliovirus. These cases were not caused by a vaccine-derived strain, but rather by a wild-type virus traced back to Uttar Pradesh in Northern India, one of the less than 20 countries where polio still occurred. The fact that the wild-type virus could reestablish itself warned of another population of children for whom susceptibility to polio had risen above safe levels.

Bulgaria had not had a case of polio in almost ten years. Europe's last case was over two years ago. The continent was only one year shy of the three necessary to certify the entire European Region polio free, a goal that everyone expected to reach at the end of 2002. No one wanted to see that goal missed.

Oblapenko and others from the WHO and CDC acted quickly. Their investigation found that both affected children belonged to the Roma, Bulgaria's gypsy population. The Roma's itinerant lifestyle places extra demands on the public health system to maintain high levels of immunity, but a campaign to vaccinate all 130,000 Roma children was quickly put into place. Mobile teams worked to deliver vaccine to those at most risk while two national immunization days were planned and then executed throughout the country.

Over 400,000 children in Bulgaria received at least two doses of vaccine over the ensuing months. The cost of the emergency effort was added to that of Haiti and the Dominican Republic. The world health authorities were now left to sit and wait to see whether new cases would appear in other areas of the world previously free of the disease. New cases would almost surely guarantee that the 2002 goal to stop transmission worldwide would be missed. In the meantime, the seconds are disappearing relentlessly off the clock at the United Nations.

The Invisible Disease

Anyone born in the Western world after 1960 is unlikely to have more than vague memories, if even that, of a disease called paralytic poliomyelitis—polio, as it came to be known. Perhaps a memory of sugar cubes from the family doctor, and of being told that these were protection against a crippling disease. Perhaps an older family member or friend or acquaintance with a withered arm, perhaps leg braces. History-book photos of FDR. Everyone has heard of polio, but an increasing number of people never have to give it a thought, and so they never do.

In this respect, these recent generations are not unlike all those up to the end of the 19th century. Before then, nobody thought much about polio. Given that the poliovirus has almost certainly been circulating among us since before recorded time, it seems strange that there should be so few accounts of the disease caused by that virus. But it's true—and not so strange, once the story unfolds—that polio became a major public health problem, and so broke into our collective consciousness, only at the beginning of the 20th century, when it reached epidemic proportions in the United States and several countries in Europe. The virus and the disease were ancient, but the epidemics were not, and for anyone living during those decades—between the 1890s and the late 1950s—it was as if some new plague had been called down.

The story of polio is a bit like an impressionist's painting, however—up close it is a confusing series of dots and color; from a distance, a clear and understandable picture comes into view. The public perceptions—of a virus that seemed to materialize suddenly, wreak its devastation for several decades, and then all but disappear—come from viewing the dots rather than the whole. As a more complete understanding of polio and its viral agent have emerged, it is clear why this particular disease is almost absent in recorded history until near the end of the 19th century. The absence ultimately tells us much about this shadowy human plague and the virus that causes it.

Medical historians agree that the earliest record of polio is the image of a young man with a shrunken leg carved into an Egyptian stele over 3,500 years ago. The young man wears the robes of a priest. A staff supports him, and his deformed limb is highly reminiscent of a modern victim of polio. A few centuries later Hippocrates wrote about a deformity he called "acquired clubfoot." From his descriptions, it seems likely that some of the cases he described were probably paralytic polio. These victims must have been relatively rare, however, as no description of an outbreak appears anywhere in his two massive volumes of medical observations, *Of the Epidemics.*

Although the art of the 17th century offers similar glimpses of polio, the first credible description doesn't appear until near the end of the 18th century. By then, precision in diagnosis had improved, and keen observation—characteristic of Hippocrates' writings—had finally begun to return to the practice of medicine. As described then, polio was clearly a rare and seemingly random disease. According to one physician, it attacked "children previously reduced by fever; seldom those under one, or more than four or five years old," but there was no discernible pattern among those it attacked.

Over the ensuing decades, medical observers and writers continued to add details to the picture: a link was perceived between the disease and exposure to unsanitary conditions; the flaccid nature of the paralysis was described; and a hypothesis was developed that the disease somehow affected the spinal cord, and in particular the nerves that power the muscles rather than the nerves that control sensation. Nonetheless, because there was still no apparent connection between victims, the disease continued to be characterized as a sporadic affliction of young children—a matter of bad luck rather than contagion.

The idea that the disease was caused by an infection and was being transmitted from person to person had to wait until the disease, or rather, its frequency, began to change. Such a change began at the end of the 19th century, with three small European outbreaks: one in Norway, another in France, and a third in Sweden. Then, in 1894, an even larger outbreak occurred in the U.S., in rural Vermont: over 130 cases in a single summer, each one carefully documented by the local physician and public health officer. The three European outbreaks and the Vermont epidemic were notable for several reasons: the cases were clustered in place and time; older children and young adults were affected; and the death rate among these older victims far exceeded the rate among younger children. Together, these factors marked a shift in the epidemiology of polio—from sporadic to epidemic, from infants to otherwise healthy older children and young adults—a pattern that held over the next five decades.

In all these observations, the notion that infantile paralysis (as it was called by then) might be caused by some infectious agent was still overlooked. Typically the blame fell on "overheating," or on fatigue or trauma, sometimes on "chilling" the body. Such explanations, emphasizing what the victim did or did not do, reflected the inability of specialists to find a link between cases, to see polio victims as part of a network of infection. Poliomyelitis was unlike smallpox or measles. The actual cause of smallpox, for example, was still unknown at this time, but if a person came down with that disease, two things at least were certain: he had caught it from someone else, and he could give it to someone else. It was relatively easy to reconstruct the movements of the disease. But polio victims just seemed to pop up, and, to deepen the mystery, they didn't seem to pose any threat to anyone else. Until an infectious agent was discovered, it would be hard to see the disease as anything other than random and self-inflicted. A person worked too hard, became overheated, and then something strange happened and the person was paralyzed.

This conception of the disease was finally overturned in 1906, when a major epidemic hit Sweden. The physician and epidemiologist Ivar Wickman investigated more than 1,000 cases of poliomyelitis, paying particular attention to the problem of how the disease moved from victim to victim. Where previous discussions of the disease had emphasized one symptom—paralysis, either temporary or permanent—Wickman saw the importance of the so-called abortive cases, those that did not lead to paralysis. Such cases had been noted before but discounted as anomalous. Wickman hypothesized exactly the opposite; most cases, he suggested, were mild, and the *usual* case of polio was a short course of fever and muscle stiffness. The disease had several faces, in other words, most of them not very frightening at all. Arguing that these flulike cases were the links between the paralytic ones, Wickman was able to track the disease as it migrated from urban center to the countryside, and thus begin to prove that there was indeed an infectious agent of some sort at work.

Meanwhile, epidemics of polio grew larger. Increasingly its victims were older and its effects more severe. Another Swedish epidemic affected more than 3,600 people. With the virus now identified, Swedish researchers were able to conduct some of the most revealing work yet during this epidemic, although their results were not universally accepted until years afterwards. In particular, researchers discovered that the virus was almost everywhere. They found it in the stool and saliva of people sick with the disease, as they expected to, but they also found it in people who had no symptoms at all. Wickman was right, in other words, but hadn't gone nearly far enough; many more were carriers of the virus than anyone had suspected, most of them perfectly healthy. They were infectious nevertheless. Furthermore, large numbers of children and young adults who had shown no sign of the disease were immune to it, thus proving that they had already been infected.

Additional outbreaks occurred in the U.S. during the first 15 years of the 20th century, and certain careful clinical investigators came to similar conclusions—this was an infectious agent far more common than originally suspected, one readily transmitted from person to person, yet for the vast majority of people infected, it was completely benign. For decades, the silence of most infections confounded the public's understanding of how the disease was spread.

Two results of this confusion were fear and misinformation. In 1916 the U.S. suffered its first massive epidemic. It occurred in the northeastern region of the country, and it was devastating. In New York City alone there were over 9,000 paralytic cases and over 2,000 deaths. Panic reigned. For the average man or woman on the street, frightened half to death, there were a thousand theories about the cause of this "new" plague. Flies, foreigners, dogs and cats, garbage, and mysterious vapors were all suspected, yet taking steps against them did no good. Summer after summer, as epidemics continued to strike around the country and as people struggled to understand what was happening to them, this same sense of panic repeated itself.

Equally remarkable were the steps taken to control the epidemic by the health authorities. Because polio had been clearly identified as an infectious disease, transmitted from person to person, public health authorities focused their attention on controlling people, particularly their movements. In New York City, epicenter of this first massive epidemic, the measures were especially draconian. All housing where an identified polio victim lived was placarded and the entire family was quarantined. Windows had to be screened, and everything was disinfected. Even the household pets were held hostage. In an extraordinary decision, travel out of the affected areas was severely restricted, particularly for children 16 years old and younger. No one was allowed to travel out of New York

City and surrounding communities during the three-month epidemic without presenting a health certificate stating that their household was free of the disease.

In retrospect, of course, such restrictions could do little good, particularly this last one. A strict control of the sick and their families won't work if healthy people, even those with health department certificates, are just as infectious—and this was the case. Needless to say, the restrictions were unpopular and were extremely difficult to enforce. Despite their ineffectiveness, however, local health authorities continued to impose similar restrictions around the country over the next decade.

The Clearing Picture

Several hypotheses from the Swedish experience were confirmed by the 1916 epidemic. First, polio was clearly identified as a purely human infection; it was transmitted from person to person in some manner, without the intermediaries of insects, animals, or mysterious vapors. Second, many more individuals were infected with the poliovirus than actually developed the disease. Third, such silent "carriers" were crucial to the spread of the disease; the frank paralytic cases, in contrast, played a relatively minor role. Finally, because of the huge numbers of people infected, even a minor epidemic served to immunize the broader community against subsequent epidemics—at least until new births and new residents once again raised the number of the uninfected.

Thanks largely to the fear and urgency these epidemics inspired, we now know a great deal more about the tiny virus that causes polio, about how it spreads and how it behaves inside a human being. The first cells the virus attacks are those that line the throat. The virus attaches, penetrates the target cell, and begins to replicate. Within a few hours, a single infectious particle has become thousands of infectious particles, so many in fact that the infected cells burst, thereby releasing the new particles. These pass through the stomach and into the bowel, where other target cells—the ones that make up the lining of the intestine—abound. The process is repeated over and over: invasion, replication, release. Because the virus attacks the throat, particles are present in phlegm and saliva, but because it thrives in the gut, millions of particles wind up in an infected person's stool.

Where crowding is intense and personal hygiene poor, the virus can rapidly make its way from person to person. Under these conditions, the virus is endemic—extremely common—and the vast majority of people encounter it as infants. As soon as the protective antibodies passed from a mother to her child decline, the child is infected. In almost all cases, the poliovirus passes through silently, leaving behind only the immunologic memory that will protect the child for life. In a small number of children, however—perhaps as few as one in 1,000 infected—the virus crosses into the bloodstream or migrates up the nerve fibers to arrive in the central nervous system. By infecting and reproducing inside the motor neurons, the virus causes neurological damage, in particular the flaccid paralysis that first prompted observers to recognize polio as a discrete disease.

As living conditions improve, however, the likelihood diminishes of encountering the poliovirus very early in life. Consequently, the number of children and young adults without immunity grows. When the virus arrives in such a community, an epidemic rapidly ensues, and for reasons that are still not clear the virus is more likely to paralyze or even kill these older victims.

This explains why epidemic polio first appeared, as if from nowhere, in Europe and the U.S.: only in the most developed nations was life sanitary enough to make an epidemic possible. The irony is that if these countries had been dirtier, if their engineers had spent less time improving waste disposal and water treatment, polio would have remained an endemic infection and a rare disease. In the wake of the improvements and the resulting epidemics, what was needed was a safe way of exposing people to the virus—a vaccine, in other words—but that would not come for decades. In the meantime, anyone paralyzed by polio, or terrorized by it, or even merely inconvenienced, was in a sense a victim of progress.

President Roosevelt's Other War

Although epidemics of paralytic polio continued to occur regularly in Europe and the United States, the battle against it became predominantly an American one. This is no doubt due primarily to the stature and the efforts of America's most famous polio victim—Franklin Delano Roosevelt. The disease reached him in August 1921, while he and his family were vacationing on Campobello Island, off the coast of Maine. He was 39 years old, the father of four children. It must have been humiliating for this vigorous, ambitious man—a Roosevelt, after all, and one clearly cut from presidential timber—to succumb to infantile paralysis. Yet after 1921 Roosevelt's legs were paralyzed forever.

Roosevelt was a symbol of hope and courage to all who suffered from the disease. He personally backed a series of efforts directed toward first caring for post-polio victims and then funding the basic research that would ultimately lead to the tools for preventing the disease. His personality and, ultimately, his presidency put him in a powerful position from which to generate wide support. In 1926 he bought a health spa in Warm Springs, Georgia, and in 1927 he established the Warm Springs Foundation. The mission of the foundation was to provide rehabilitation and care for victims of the disease and reflected Roosevelt's belief that the effects of polio could be cured by a proper regimen of exercise and diet. Unfortunately, this was not so, but his success with Warm Springs no doubt helped inspire a second initiative, which would turn out to have a major impact.

Roosevelt went on to establish The National Foundation for Infantile Paralysis (NFIP) in 1938. The NFIP was a unique partnership of scientists and volunteers. It was the first organization created to attack a single disease, raising money and funneling it into care for victims and into research and development. In the foundation's first grassroots effort at fund raising, it asked the public to send dimes to President Roosevelt at the White House, an effort that soon became known as the March of Dimes. Between its founding in 1938 and 1962, when the second polio vaccine was tested and approved, the NFIP raised and distributed over $US 630 million. While it would be naïve to claim that the NFIP itself solved the problem of polio, the money it made available was vital to that solution.

As the research money began to flow, important discoveries followed, perhaps the single most important of which occurred in 1948. That year, three researchers at Children's Hospital in Boston—Drs. John Enders, Thomas Weller, and Frederick Robbins—found a better way to propagate the poliovirus in the laboratory. Prior to this breakthrough, it was hard to see how researchers would ever be able to grow enough poliovirus to mass-produce a vaccine. Scientists had been able to propagate the virus in only one kind of test-tube medium: cells from the nervous system. These cells were not only hard to get, they were also unsafe for vaccine production. The virus could be propagated in live monkeys, but this was equally unsatisfactory—monkeys were expensive, and recovering enough virus from them for vaccine production was virtually impossible.

Using improved tissue culture techniques, however, Enders and his colleagues proved that the poliovirus would grow in non-nervous system cells. The effect of this discovery was to bring the monkey and the test tube together. Now a single monkey kidney could provide hundreds of tubes of culture medium, a plentiful, convenient, and inexpensive source of the virus. The achievement opened the way for many of the discoveries and developments that followed. It was so important, in fact, that it won Enders and his colleagues a Nobel Prize in Medicine.

The next major advance came four years later. For decades researchers had known that infection with the poliovirus conveyed immunity to subsequent infection. No one knew, however, how many different strains of the virus existed, and it was possible, as with influenza, that immunity to one type of poliovirus did not mean immunity to other types. Because the ideal vaccine should protect against all strains, it was important to discover exactly how many strains there were. The ability to grow the virus in the laboratory helped to speed up this vital work, and by 1952 the research community was able to say for certain that there were only three types, opening the door for the rapid developments that followed.

Other findings brought researchers closer to a solution. They discovered that once the virus established itself in the gastrointestinal tract, it invaded the bloodstream. This made the virus more dangerous—it was through the bloodstream that the virus made its way to the spinal cord—but it also made it more vulnerable. The immune system, stimulated by vaccination, could attack the virus much more efficiently once it entered the bloodstream.

By now, the NFIP had begun to see its role as not only chief fundraiser in the fight against polio, but chief strategist as well. Armed with the findings, the foundation's leadership moved aggressively forward to develop the ultimate weapon to "conquer polio." Its choices determined the fate of a waiting and fearful public.

Ethical concerns immediately began to conflict with the science of vaccine development, and the NFIP sat decidedly in the middle of these debates. The most vexing issue involved human experimentation. Revelations from the Nuremberg trials about medical research in Nazi Germany had tainted the very idea of using human subjects, for scientists and the public alike. Yet it was clear that defeating polio would require, at some point, testing candidate vaccines on humans, and in particular on children, because they were at highest risk for the disease. Eventually the question shifted from whether to use children to how to use them and when, but this didn't make the NFIP's position any easier. The foundation found itself caught between a public eager for a solution to the polio problem and a scientific community committed to slow and painstaking work.

Among scientists, one question predominated: should the vaccine be made from killed virus, as influenza vaccines were, or from a live virus, attenuated in its virulence, like the measles vaccine? Each had advantages and disadvantages, and many of these became increasingly important as the world moved toward global polio control.

One advantage of the killed-virus vaccine was that it was almost ready for trial. By 1953 Jonas Salk, who had led the effort to determine the number of polio types, was ready to begin testing his killed vaccine. Plus, a killed vaccine raised fewer concerns about safety. Salk's vaccine might or might not work, but there was little chance that anyone receiving it would get a virulent dose of the virus. On the negative side, the killed vaccine was cumbersome. It had to be injected, and not once but four times, in order to achieve solid protection against the virus. This took months and required special equipment, not to mention special personnel—a trained health care worker at the very least. Another drawback concerned the type of immunity conferred. The poliovirus proliferates in the gut, its natural habitat. But the killed vaccine, because it is injected, can stimulate immunity only in the bloodstream. This means that a person vaccinated with the killed vaccine—the Salk vaccine, as it came to be called—would be protected from the disease but might still be infected by the virus. Such a person might still be able to transmit the virus to others.

With the live, attenuated polio vaccine, the positives and negatives were reversed. It was still several years away from being ready for trial, and a real danger was inherent in it: because it was still alive, the vaccine virus might revert through mutation to full virulence. Assuming that this danger could be eliminated, however, the live vaccine had much to recommend it. It was an oral vaccine and therefore wouldn't require trained personnel or special equipment; one could give it on a sugar cube. Further, it would produce a faster and longer lasting immunity because the attenuated particles would act in the body very much like the native virus, setting up an infection, and therefore producing immunity in the gut.

This type of immunity has two advantages over the type of immunity produced by the killed vaccine. First, there is reduced danger that the person with gut immunity will later contract the wild-type virus and spread it silently. On the contrary, the live-virus vaccine sets up a kind of virus factory in the gut, turning out millions of copies of the attenuated strain. The vaccinated person sheds these particles, just as he or she would the native, wild-type virus, and anyone who ingests them gets an immunologic boost without even knowing it. This is the second advantage.

The debate over the merits of these vaccines was intense, but as public pressure mounted for a solution to the polio problem—1952 was the worst year yet, with over 20,000 paralytic cases and 3,000 deaths—the NFIP overcame its ethical concerns about testing on children and threw its weight and resources behind the killed vaccine strategy. By December of 1953, a massive trial of Salk's killed-virus vaccine was ready to launch.

The NFIP announced the results of the Salk vaccine trial on April 12, 1955, the tenth anniversary of Roosevelt's death. The trial involved over 1.8 million children and was completed in record time for such a large trial. Further, it was an extraordinary success. It was clear from the results that the vaccine worked as intended. Amid intense media coverage, representatives of the U.S. Public Health Services made the decision on the spot to license the vaccine. Although such speed in decision making was (and remains) antithetical to the orderly pace of scientific review and public policy formulation, public desire for a vaccine was so intense that to act otherwise—to have said, in essence, "Let's wait a year or two or three, just to be certain"—would have been extraordinarily difficult.

As it turned out, the decision was the right one. Except for one disastrous incident in which a manufacturer supplied a batch of defective vaccine containing living polioviruses, the Salk vaccine was not only safe but also immediately successful. By May 7, 1955, over four million doses had been administered in the United States alone. Canada and Denmark began immunizations immediately as well, and use of the vaccine spread rapidly throughout the developed world. By 1962, Sweden was polio free, and the number of cases of paralytic polio in the United States had dropped from 20,000 per year to just over 100.

The Needs of the World

Given that the Salk vaccine was both a major medical and public health victory, it's easy to understand why the NFIP was pleased with it, and with itself, and why some of its leaders tended to overlook the needs of the wider world. Nevertheless, while the industrialized world celebrated, things went on much as usual in the developing world, where the poliovirus remained endemic and the threat of paralytic polio was constant. In the poorest regions of the poorest countries, the Salk vaccine was only slightly more useful than an X-ray machine. Full immunity required four injections. That may not sound like much, and for most children in Akron, Ohio, or London, England, it wasn't. But in a country without clinics, equipment, trained personnel, or even familiarity with Western medicine, four injections might as well be 400.

The answer to this problem lay in the research the NFIP had turned away from in the early 1950s: a live-virus vaccine, one that could be taken orally. In the 1960s the World Health Organization began to advocate this approach, and in so doing took the first small steps toward a global polio eradication plan.

In settling on the Salk killed vaccine, the NFIP had set aside the advice and work of several important researchers who were pursuing an attenuated vaccine. Chief among these was Albert Sabin. Fortunately, the force of his personality, not to mention the quality of his research, was such that he continued his work even without support from the NFIP. He had been one of the vigorous advocates of a live vaccine within the NFIP's advisory committee, recognizing that a killed vaccine would almost certainly be useful only to children in developed countries.

In 1957, two years after the Salk vaccine was approved, Sabin had his attenuated vaccine ready for human trials but found himself unable to test it in the U.S. The distribution of the Salk vaccine had been so effective that not enough unvaccinated people were left in the U.S. for a second massive trial. Sabin turned to colleagues in the U.S.S.R., where interest in a live vaccine was high. In spite of the Cold War, a collaboration was born. Soviet scientists were able to conduct large-scale field trials that proved both the safety and the efficacy of the vaccine and cleared the way for its use throughout the world.

The number of polio cases had already fallen dramatically in the industrialized world, and with the appearance of the Sabin vaccine, the number fell further. Any hope, however, that control would spread as rapidly in the developing world was soon dashed. The availability of a simple weapon against the disease was not enough to prompt its use on a global scale. In addition to the usual challenges of limited resources and inadequate or nonexistent health care infrastructure, mounting a global campaign against polio presented unique difficulties. Unlike the smallpox vaccine, which could be freeze-dried and carried into the remotest areas without losing its potency, the oral polio vaccine is subject to rapid deterioration. If the vaccine warms up, the live viruses lose their infectivity and thus their ability to generate immunity. Anyone who wanted to take the vaccine into the developing world had to find some reliable way to keep it cold.

Surveillance presented a double challenge. By the late 1960s it was understood that surveillance—the ability to know, quickly and accurately, where the active infections were—was an indispensable part of any control strategy. Against some diseases surveillance is relatively easy because the infectious agent can't hide. To be infected with the smallpox virus, for example, is to develop the disease smallpox, and even the mildest case is obvious on the skin of the infected person. The poliovirus, in contrast, is much sneakier: the majority of people infected with it suffer no illness at all. These active infections are impossible to spot, which means that the virus can circulate in a community for a long time before it finally manifests itself, setting up the conditions for a full-scale epidemic.

Further, even when the virus has caused disease, distinguishing it from other viruses remains difficult. Mild cases of polio might look like the flu. Several different viruses can cause flaccid paralysis, the hallmark of the most serious form of poliomyelitis. The only way to specify the disease agent is to recover it in the laboratory, an expensive and relatively slow proposition. Confirmation of cases thus depends both on well-trained health care workers and on sophisticated laboratory services, neither of which were common in developing countries.

Faced with such obstacles, the drive first toward worldwide control and then toward eradication had to await the leadership of people and organizations with goals of a very special nature. These didn't come together until the early 1970s, almost a decade after the oral vaccine became available. A nationwide campaign in Brazil served as a testing ground for the efforts against polio. The campaign mounted there showed how much could be accomplished, even in a developing country, when the resources and political and social will could be brought together. Success in Brazil began to convince some that global eradication of polio might be possible.

Brazil's Attack

Brazil typifies many of the challenges inherent in mounting a major campaign against an infectious disease in a developing country. Through most of the 1960s, the live polio vaccine was available through the Ministry of Health, but supplies were erratic and no national program promoted its use. The unstable vaccine had to be refrigerated during transportation and storage in a so-called cold chain from factory to vaccinee, yet electrical services were unreliable in many areas of Brazil, and many local clinics didn't have refrigerators even when power was available. The Health Ministry's organizational structure was ineffectual, with most of the administrative decisions carried out at the state level, and the local health care systems were either inadequate, nonexistent, or dominated by physicians more interested in wealthy patients and their expensive diseases than in public health matters. In short, there was neither political nor social commitment to the idea of making the new vaccine available to everyone. As a result, only the children of the wealthiest families were vaccinated. The poor made do as they had always done, and so did the poliovirus. Each year there were over 7,000 cases of acute flaccid paralysis, and the virus circulated freely.

This picture began to change in 1971 when the Ministry of Health launched the National Polio Control Program, the first attempt to provide the Sabin live oral vaccine to preschool children throughout Brazil. The basic idea was a good one. The ministry would provide the vaccine but encourage each state to design its own strategy for mass immunization.

The virtues of the oral vaccine were such that brief, intense campaigns should have been viable. Because of the ease of administering the oral vaccine, it could be given by volunteers rather than trained health care workers. This expanded the pool of people available to administer the vaccine, making it possible to conduct large campaigns in only a few days. Short but intense immunization campaigns could also take advantage of the fact that the vaccine strain circulates through a community just as the wild poliovirus does—along the fecal–oral route—and thus creates collateral immunity among any children still unvaccinated.

Unfortunately, it soon became clear that relying on each state to organize its own program yielded less than satisfactory results. The initial response to the National Polio Control Program was reasonably enthusiastic, especially among poorer states, where fewer health services meant lower levels of immunity. Not all states, however, organized the mass campaigns envisioned by the ministry, and even those that did lost interest after the first year. Although surveillance was poor and numbers are therefore unreliable, it appears that the incidence of polio was unchanged: approximately 7,500 cases of paralytic polio each year.

In 1974, a different issue presented another kind of challenge. The narrow agenda of the NPCP—polio immunization and nothing else—conflicted with the broader public health mandate of the newly appointed political leaders of the Health Ministry to increase the quantity and improve the quality of primary and preventive health care for everyone. For many public health specialists, the notion of a massive campaign against a single disease compromised the public's interest in routine prevention practices, plus scarce resources meant that one would be funded at the expense of the other. The arguments for and against the two approaches—sometimes called vertical versus horizontal—were repeated again and again, not only in Brazil but also within the WHO itself. The result in Brazil was that the Ministry of Health incorporated the polio campaign into its larger National Immunization Program designed to provide several vaccines that could be administered through routine health services.

The one benefit to come out of the NPCP was a strengthening of the surveillance system; this system now included several regional laboratories prepared to isolate the virus and thus confirm suspected cases. Otherwise, outbreaks of polio continued unabated, and vaccine coverage among the poorest parts of Brazilian society dropped as low as 10%.

As perhaps is too often the case, it took an outbreak among the wealthier states in Brazil to prompt a change in strategy. In December 1979, a major epidemic of polio took place in the states of Paraná and Santa Catarina. Fortunately for Brazil, the epidemic coincided with the appointment of João Baptista Risi, a public health specialist schooled in the successful smallpox eradication campaign, as National Secretary for Basic Health Activities. Risi had a realistic understanding of the value of nationally coordinated and targeted programs, for he had participated in the NPCP. Armed with knowledge of that program's failings and with the strength of his convictions, over the next 14 years Risi led Brazil's transition from a polio-endemic to a polio-free country. This victory was important not only for Brazil but for the rest of the world as well because it proved that transmission of the virus could be interrupted even in a developing country.

Risi argued for a mass vaccination campaign, organized at the national level. His ambitious goal was to orchestrate the immunization of all Brazil's preschool children on a single day. Although the logistics of this were a considerable challenge, he won the highest level of support—that of Brazil's president. The first "National Vaccination Day" was set for Saturday, June 14, 1980. Risi commandeered the state-owned slots for public announcements on the various media, using 123 television stations, 1,200 radio stations, and 3,900 loudspeaker services for 15 days prior to the event.

As June 14 approached, no one knew whether the elaborate plans could be carried out or whether the public would respond to the call for vaccination. Tension among the organizers was high. Risi recalls, "The experience of D-Day . . . was an unforgettable one. There was no way to insure appropriate strategic supplies of vaccines, since one could not predict the demand at individual sites. . . . Weather conditions were another reason for concern. . . . Most importantly, there was some uncertainty about the degree of public responsiveness."

As it turned out, the public was responsive indeed: by the end of the day, 17.5 million children had received the oral polio vaccine. The day was an extraordinary achievement, involving 90,000 distribution sites, thousands of tons of ice, and more than 300,000 volunteers. Though the success was a relief to Risi and his colleagues, questions remained. How sustainable would the undertaking be? One dose of vaccine was not enough. Children needed a second, and then a third. The next National Vaccination Day, scheduled for two months later, would benefit organizationally from the experiences of the first, but no one knew whether the public would return, as the reason for a second dose of vaccine was probably not clear to most.

Nevertheless, the second day was as successful as the first, and Brazil's program to control polio was under way. Within the next three years, polio cases throughout Brazil dropped from 7,500 per year to only 45 confirmed cases in 1983. As important as this was for the country, it was even more important for the world. National immunization days were not new. They had been used successfully in Cuba, but Cuba is an island and has a totalitarian government. Brazil was the first time the strategy had been applied in a developing country on the South American continent, one lacking the level of government control present in Cuba. Risi had proved to all that large-scale immunization drives could be mounted even in countries where health care resources were limited.

Coming to America

The idea that polio could be eradicated found an early supporter in another Brazilian who had worked in the smallpox campaign. Ciro deQuadros watched Risi's results in his native country carefully, and as early as 1981 he became convinced that another global eradication effort would pay off. For some time, both Salk and Sabin had believed that their vaccines offered the necessary tools, and each had weighed in with the opinion that eradication was now viable. Both of them found their champion in Ciro deQuadros.

In the early 1980s, however, a global campaign to eradicate polio was not the easy sell that some might have expected it to be, and resistance came from some surprising quarters. In spite of the recent triumph of the smallpox eradication program, the WHO displayed almost no enthusiasm for another campaign of a similar nature. Behind this resistance stood the now familiar tension between vertical and horizontal approaches to public health—massive, targeted campaigns on one side and general health care on the other. In 1978, the WHO had held a major conference in Alma Alta, Kazakhstan, on the need for primary health care.

The result was a significant commitment on the part of the world's health ministers to improving primary health care for all citizens. The WHO firmly held the belief that another eradication campaign would take attention and resources away from this larger goal. DeQuadros recalls the director-general of WHO, Halfdan Mahler, telling him that "never again would such a vertical program [smallpox eradication] be promoted by the World Health Organization." The debate between those who favor targeting a single disease and those who favor a general improvement in health care has continued for the ensuing decades.

Resistance to a polio eradication campaign also rested on biological arguments. D. A. Henderson, for example, who had led the Smallpox Eradication Programme and who understood the value of vertical initiatives, was skeptical. His reaction to deQuadros's suggestion was characteristically blunt: "No way!" In his estimation, the poliovirus was simply too difficult an opponent. "Even with smallpox, there were just so many things that could have gone wrong, and smallpox was a much easier disease to tackle."

Still, events over the next two years made the ground increasingly fertile for deQuadros's ideas. The Pan American Health Organization (PAHO, the WHO's regional office for the Americas) elected a new director, Carlyle Guerra de Macedo, in 1983. Macedo was keenly interested in improving childhood immunization, an issue that was also a high priority for UNICEF's director, Jim Grant. Thus, both organizations' leaders were important advocates for universal childhood immunization, a WHO goal set for 1990. When questioned about the feasibility of reaching that goal in the Americas, deQuadros saw his opportunity to place the eradication of polio back on the agenda.

DeQuadros maintained that if Macedo and Grant wanted to renew interest in an immunization program, they needed a "banner disease." There was none better than polio, in part because the U.S., Canada, and several South American countries were already well on their way to controlling it. Besides, a campaign to eradicate polio could be structured so as to support the distribution of other vaccines, improve disease surveillance, and generally promote the goals of improved primary health care espoused by the WHO. In other words, the campaign could be structured so that it served both vertical and horizontal goals. His arguments were persuasive, and thus a full-fledged polio eradication initiative backed by PAHO began in the Americas in 1985, with deQuadros as director. Of equal importance, Jim Grant and UNICEF became strong supporters of polio eradication.

The polio campaign required a different strategy from the smallpox eradication program. For smallpox, it became clear early on that mass vaccination was unnecessary; in this respect smallpox was, as Henderson suggested, relatively easy to fight. Smallpox was not readily transmitted from person to person, and thus it was necessary only to find out who had the disease and then vaccinate everyone in contact with that person. This was not the case with polio. Because of polio's distinct biologic properties, it would be necessary to immunize 90% of the population at highest risk—children between the ages of six months and five years—and to maintain that level of immunization until the disease was eradicated. And ultimately, because imported cases were always a threat, true eradication would require a global strategy. Control in a single region or continent, no matter how complete, could not last long if the disease was wild elsewhere. This posed a daunting challenge.

The Brazilian experience reinforced two important issues for PAHO as it planned for eradication in the Americas. First, routine immunization services in developing countries were not sufficient to generate the levels of polio vaccine coverage necessary to interrupt transmission. The requirement that 90% of newborns receive three doses of oral vaccine simply could not be met, even if there were enough vaccine available. A different mechanism for vaccine delivery had to be used. Second, individual states, no matter what the final plan might be, could not be relied upon to design and coordinate a successful campaign. Coordination needed to happen at the national level at least.

The design of the campaign in Brazil was the key. Supplementing routine immunization services with national immunization days had rapidly brought the disease under control; thus, national immunization days became the accepted strategy for polio eradication. Under deQuadros's leadership, each country in the Americas in which polio was still endemic was asked to develop its own delivery plan, but the essential strategy was the same—give as many children as possible, five years old and younger, a dose of vaccine on a single day, regardless of immunization status, and then repeat the event within a month. This strategy would need to continue on at least an annual basis until the number of cases had dropped significantly, at which point a mop-up campaign could be used to break the final chains of transmission within a region.

This effort carried a big price tag. Although the oral vaccine itself was inexpensive, mounting national immunization days, improving surveillance systems, and developing the laboratory services necessary to support the plan were not. PAHO estimated that the campaign would require at least $US 90 million more than the organization had available. This money came, surprisingly, from a new player on the stage of international health.

It was fortunate for PAHO, and for victims of polio, that Rotary International, an organization with branches in almost every country and a long and distinguished record of community service, was looking for a global goal to mark its centennial celebration in 2005. Albert Sabin, developer of the oral polio vaccine, suggested to Rotary International's president that it might be possible to eradicate polio. The idea was an almost immediate success. This was a project with global impact, and it suited the organization's mandate, skills, and resources. In 1985, Rotary International began Polio Plus, the largest humanitarian effort ever undertaken by a private organization. Since then, Rotary International has contributed close to $US 500 million, mobilized hundreds of thousands of volunteers, and brought to bear significant political pressure, not only in the Americas, but throughout the world.

Several government agencies soon followed Rotary's lead. UNICEF, the Inter-American Development Bank, and USAID all contributed money and expertise, and an interagency coordinating committee was put into place. The result was a first in the history of global public health: a massive, coordinated effort of both public and private organizations. It is a tribute to deQuadros's skill that these partners worked together so well. Once efforts got under way, the number of polio cases plummeted. Perhaps because polio is a disease of children, the level of commitment—social, political, and financial—was extraordinary, so much so that even civil war could be interrupted, at least temporarily. In El Salvador, for example, PAHO, UNICEF, the Red Cross, and the Catholic Church all took a hand in negotiating "days of tranquility," during which government workers and guerrillas worked together to deliver vaccine to the nation's children. Efforts like these, many hoped, would foster improved understanding and mutual purpose, becoming the basis for "Health as a Bridge for Peace."

The polio campaign had the power to bring people together even under the most challenging circumstances. In Peru, guerrilla actions by the Shining Path were a constant threat to the polio campaign. Unlike El Salvador, the Shining Path had no centralized organization with which to negotiate, and guerrilla units could and often did disrupt government activities. When the final push to eradicate polio came in Peru, PAHO wanted to immunize children from almost two million households, essentially the entire country, in a single week—a massive action that offered the guerrillas a perfect opportunity for disruption. Concern for the nation's children prevailed, however, and the week proceeded without incident. A few months later, the government mounted a major campaign against the Shining Path, arresting a number of its most important leaders and guerrilla fighters. Much to everyone's surprise, several of those arrested had been volunteers during the weeklong polio vaccination effort.

Peru typifies the community spirit that surrounded the polio campaign in other ways as well. In the final push, over 11,000 volunteers from Rotary International participated. A staggering amount of ice was needed in order to maintain the cold chain for the vaccine. Lima, for example, needed close to ten tons of ice. Throughout the country, volunteers went from house to house every night, using people's refrigerators to freeze the cold packs and ice trays needed for the next day's vaccinations.

Peru was also the country where, in 1991, public health officials found the last case of polio in the Americas—a two-year-old named Luis. In August 1994, the Western Hemisphere was declared free of polio.

Going Global

Ciro deQuadros had the vision and the courage to mobilize an entire region, test the strategies, and prove that they worked. The PAHO program was so successful so quickly, in fact, that even before it was finished it had convinced many public health officials that the time had come for another ambitious goal—the global eradication of polio. This belief, coupled with the extraordinary financial commitment and the political pressure mounted by Rotary International, finally brought the WHO on board.

The remarkable success in the Americas had a strong influence on the agenda for a seminal meeting held in Talloires, France, in March 1988. Bill Foege, another veteran of the smallpox eradication campaign, had organized the meeting under the auspices of The Task Force for Child Survival and Development. Risi and deQuadros were among the 60 attendees, and they made the case for polio eradication. Jim Grant continued his support as well. Halfdan Mahler, the director-general of WHO, who had carried the banner against vertical programs for over ten years, changed his position at the meeting; he left convinced that polio eradication could, in fact, help build primary health services. Two months later, Mahler placed polio eradication on the agenda for the World Health Assembly. All 166 member states attending the 41st World Health Assembly in May 1988, launched the Global Polio Eradication Initiative.

Timing is everything in life, however, and the timing was not good for polio eradication in 1988, even though things were going well in the Americas. Despite WHO's success against smallpox less than a decade before, and despite its bold declaration of intent, a great deal of inertia and skepticism still existed within the organization. Some argued from a biologic basis: the eradication of polio simply was not possible. Others, thinking more in terms of politics and policy, felt a polio campaign wasn't desirable. In 1988, the WHO was nearing the objectives of the Extended Programme on Immunization—an ambitious drive to raise universal childhood immunization coverage for six diseases, including polio, to 80%. Some worried that a program focused on polio might endanger that goal. Beyond this specific concern were the familiar doubts about vertical programs in general, with prominent voices arguing that a targeted program was not an appropriate use of resources and that developing and improving health systems horizontally was the best way to advance world health.

Slowly, however, individual countries and then entire WHO regions began to implement the proven strategies used by PAHO in the Americas. In 1993, China proposed to begin conducting national immunization days. During its first NID, volunteers and health care workers distributed vaccine to over 80 million children. Other countries in the WHO's Western Pacific Region followed suit, and under the extraordinary leadership of the Western Pacific Regional Office, the number of cases in the region plummeted. In March 1997, a two-year-old girl in Cambodia became the last case of polio in the region. In 2000, the region was certified free from polio.

By 1995, national immunization days were being conducted in 51 countries, and nearly 300 million children—almost half the world's children under five years old—had received at least two doses of oral polio vaccine. The Southeast Asian Region, representing some of the most densely populated countries in the world, including India, began to gear up. For the first time, countries began to coordinate their national immunization days, which allowed better coverage of migrating populations. In 1996, 26 sub-Saharan countries held coordinated national immunization days, immunizing over 50 million children. Similar events occurred throughout the other WHO regions, and the number of cases continued to drop.

As the pace of the campaign increased, additional donors joined Rotary International and the CDC in the effort—notably, the Bill and Melinda Gates Foundation, the United Nations Foundation, the Organization of Petroleum Exporting Countries (OPEC) Foundation, the European Union, and the World Bank—pouring resources into the campaign.

Of equal importance, the list of nongovernment organizations (NGOs) grew. Organizations such as International Red Cross, Red Crescent movement, and Médecins Sans Frontières were critical in helping to gain access to children in hard-to-reach areas and conflict-affected countries.

By 1998, just ten years after the WHO began the campaign, only 6,349 of the world's children contracted paralytic polio—less than 5% of the estimated 350,000 cases in 1988. But 1998 was only two years away from the target date set for stopping *all* transmission of the virus, and the planners knew they were in trouble.

Measures of Success—or Failure

Since its disappointing attempt to eradicate malaria, WHO has developed much more sophisticated indicators of whether a campaign is succeeding or failing. For polio, two objectives, if reached, would signal that the program had entered its final phase. The first of these was the number of children worldwide under five years old who had received at least three doses of vaccine, three being the minimum number required to ensure immunity. Most believed that it would be impossible to stop transmission of the virus unless at least 80% of these children were vaccinated. Monitoring conducted in countries where the virus was still circulating indicated that some were not achieving this objective.

The second objective was to ensure the quality of the surveillance information being reported to officials at WHO. As had been shown during the smallpox campaign, surveillance is important to every phase of an eradication program, but especially so as the end approaches and the strategy changes to one of mopping up—that is, surrounding every case with mass immunization. These last cases have to be identified before the mopping up can occur. Further, in order to convince the WHO and the world at large that polio transmission had indeed been stopped, there had to be convincing evidence that cases of polio *could* be identified.

Two aspects of this objective deserve comment. On the one hand, eradication officials needed to know that the surveillance networks were doing their jobs, that they were actively searching out cases. On the other, they needed to know how many of the cases reported actually were cases of polio. In response to the first need, the eradication program's technical advisors hit on a clever way to monitor the monitors: they checked the number of cases of acute flaccid paralysis (AFP) reported by each country. Because several viruses in addition to the poliovirus cause AFP, it is a constant presence even in polio-free countries;

about one person in every 100,000 develops it. WHO officials used this figure as the minimum standard, reasoning that there ought to be *at least* that many cases of AFP in a polio-endemic country. If a country reported less than one case per 100,000, officials knew that the surveillance teams weren't looking hard enough.

But even if an appropriate number of cases were being reported, the second problem was to determine which of these were polio and which weren't. This could be done only in a laboratory, so WHO set a verification standard of 80%. Eight out of every ten cases reported had to be confirmed in a lab, using two stool samples taken 24 hours apart. This was a high standard, particularly for poorer countries, and it required the development of WHO-accredited laboratories throughout the endemic world.

These surveillance standards helped to identify problem areas around the globe. Often the regions that needed the most help were the ones that claimed to need the least. In 1998, for example, the 48 countries in the African Region reported only 993 cases of polio. On the surface, these numbers indicated extraordinary progress against the disease. On closer inspection, however, the numbers raised more concern than hope. The African Region had reported only 0.3 cases of acute flaccid paralysis for every 100,000 children; of these, only 35% had been confirmed in a laboratory. The report of 993 cases was thus doubly suspect—both too high (the majority of these cases might not even be polio) and too low (many actual cases probably had gone undetected). From this analysis, WHO officials learned where they needed to concentrate their efforts.

The region that continued to record the highest number of polio cases in 1998 was Southeast Asia, with 4,775 of the 6,349 cases in the world. Over 90% of these were in India. While a few countries such as Nepal and Bangladesh were obviously struggling with surveillance requirements, this was not the case for India; it had one of the best surveillance infrastructures of all the developing countries. Its massive population, however, coupled with its areas of extreme poverty, was a tremendous challenge. There were 25 million children born in India every year, and reaching them with sufficient doses of vaccine was almost impossible.

By October 1998, it was evident that WHO was going to miss its goal: it would not stop transmission of polio by the year 2000. The program was now facing its greatest challenges. Most of the remaining polio-endemic countries were the poorest of the poor, often in civil conflict, and they lacked the medical infrastructure to finish the job. India and Bangladesh were faced with their enormous population growth as well.

In a quiet office in Building 12 of the CDC's campus in Atlanta, a woman sat contemplating this dilemma. Linda Quick was the child of missionary parents, so her interest in international health grew from deep roots. Trained as a pediatrician, she had spent all her professional years working in developing countries. Although she had not worked with the polio eradication program, she was no stranger to the challenges it faced because she had come to the CDC after working two years providing care to refugees in Bosnia. Within the CDC's Global Immunization Division she found others like herself—people with vision, enthusiasm, energy, and the ability to persevere even under the most challenging circumstances. She was assigned to the polio unit, the CDC group designated to provide technical and logistical support to WHO during the eradication program.

Quick's response to the challenges facing the eradication program was a straightforward one. If the Polio Eradication Initiative were going to succeed, it had to get more people into the field to support the countries that were lagging behind. She had her own ideas about how this could best be accomplished, ideas that were validated when she met with some of the "smallpox warriors," those people still left at the CDC who had hunted down the smallpox virus to its last places on earth. As a result, she proposed a program that would come to be named STOP (Stop Transmission of Polio).

STOP was designed to provide a key ingredient—human resources on the ground—as the number of cases diminished and the battle became more difficult. The basic idea was to send health care professionals with strong international experience to work with a country counterpart, not at the national level, in an office at the health ministry, but at the district level. For prospective volunteers, a particular area of expertise mattered less than did an understanding of epidemiology and a commitment to public health. What was essential was an ability to work well with people, especially from other cultures, and an enthusiasm for the effort ahead. As Quick explains it, "I had a fundamental belief that if you sent great people, regardless of their specific health care backgrounds, they would do great things."

The people who could make decisions quickly saw the merits of her plan. Quick believes that her naïveté may have helped her avoid most of the bureaucratic hurdles, but it's more likely that her idea was so appealing that it was easier to cut through the normal red tape. By January

1999, only three months after her STOP proposal was accepted, the first team of 23 volunteers was trained and dispatched into five countries—Bangladesh, Burkina Faso, Nepal, Nigeria, and Yemen.

Each country had slightly different needs, and the STOP team members adapted. In some countries, like Nepal, there was no surveillance system in place. In others, like Bangladesh, the surveillance system wasn't performing up to the international standards demanded by the eradication effort. In every country, regional surveillance officers were operating with little positive feedback from their supervisors, and in many instances the field staff was simply burned out. After mounting first one successful national immunization day, and then another and another, many staff members simply didn't have the energy, on their own, to do it again.

The first team was a major success, despite the confusion and disorder that often accompany a new international program. All 23 volunteers returned believing that the program could provide an important surge of energy during the final push for eradication. Their country counterparts mirrored their enthusiasm. Within the first year, countries were actively seeking STOP teams to help with everything from the planning and monitoring of national immunization days to the troubleshooting of surveillance programs.

Over the next three years, STOP teams continued to go out to the countries in greatest need. Although volunteers from the CDC made up the first team, the next seven teams, while dominated by U.S. volunteers, included health care professionals from 23 other countries. Their composition reflected diversity, from the young Epidemic Intelligence Service officer completing her training to the 71-year-old Rotarian physician who still wanted to make a difference. In contrast to the smallpox warriors, at least half of the STOP team members were women.

By fall 2001, 286 volunteers had traveled to 22 countries, often living and working in extremely challenging circumstances. They and their country counterparts by necessity went to areas where conflict and poverty presented high personal risks, because it was in such places that the need was greatest. Workers carried on in spite of diarrhea, broken equipment, rain, and heat. Two men doing house-to-house monitoring in a remote province of Pakistan were arrested when the women in the neighborhood called the police. Another had to be evacuated along with other WHO workers from Afghanistan after representatives of the Taliban government came to search their compound. But they made a difference, and the number of cases continued to fall. As the summer of 2001 drew to a close, surveillance systems around the world were functioning adequately, and immunization rates worldwide were approaching 80%. The case count was at 367, and the wild-type virus was endemic in only 14 countries.

Then world events took over. Terrorists struck the World Trade Center in New York City and the Pentagon in Washington, D.C. In the words of many, the world changed. STOP 9, scheduled to begin on September 16, 2001, had to be canceled. It was the largest team yet assembled, more than 40 people willing to set their lives aside for three months to push the virus into a tinier and tinier corner. Quick remembers how hard it was to make the decision. "This was the first time we canceled. It was a very hard decision, but we had people from all over the world being sent all over the world. Security was just too great an issue. We couldn't put the team members in a situation with that much risk."

The coordinated National Immunization Day that had been planned for Pakistan and Afghanistan in mid-September still took place, but without the help and supervision of the WHO and UNICEF staff, all of whom had been evacuated shortly after the attacks in the United States. Civil unrest continued to mount in Sudan, Nigeria, and Somalia—three of the 14 countries in which the virus still circulates. In the opinion of most, the eradication campaign will most certainly be delayed further by world events. A month after the attacks, Steve Cochi, who has led the CDC's part of the polio eradication effort, reflected on the setback. "It's a frightening thought that we won't achieve [eradication] after coming so close. I think there's more optimism than ever before that the goal can be achieved because there's been such consistent progress even in the most difficult parts of the world. So I have optimism that we will reach the goal; it's just a question of when."

In the meantime, the people who have invested so much in the largest public health initiative ever attempted watch the clock.

The End Game

As complex as the eradication initiative has been, and as vulnerable to political events, most public health people believe that transmission of the polio virus will be halted. But another serious challenge awaits the organizers of this campaign: global immunization policy after eradication.

There are two basic options, and neither is ideal. One option is to continue immunization for at least some period after transmission has been stopped. But officials have to weigh the biological risks of continuing to immunize, at least with the oral vaccine, carefully. The complications associated with its use are well known. The vaccine strains can revert to virulence, causing paralytic poliomyelitis. This event is rare, occurring in only one out of every two million or so children vaccinated, but the consequence is tragic. Concerns about this potential for reversion are such that many industrialized countries, the U.S. included, have switched from using the oral to the killed vaccine.

Until recently, however, vaccine-derived poliomyelitis remained an isolated incident. In other words, the revertant virus didn't spread and cause disease in other people. As the 2000 events in Haiti and the Dominican Republic proved, revertant vaccine strains can also be the source of an outbreak. In fact, the revertant vaccine strain probably circulated undetected among the residents on the island for as long as two years. During that time, the number of people not immunized continued to grow—an out-of-sight-out-of-mind reaction to the disappearance of disease from the Americas in the early 1990s. As soon as a large enough group of unvaccinated children existed, the virus struck. The epidemic was small, affecting only 21 children, but a similar event in a populous country like India or Bangladesh would be much more serious.

Consequently, many people involved with the campaign at both CDC and the WHO believe that it is essential to stop using the oral vaccine as soon as possible. Walter Dowdle, who serves as a senior consultant to the WHO Polio Initiative, states the issue in no uncertain terms: "You cannot say that you have done your job until you can stop using OPV, because until you stop using OPV, you're going to have vaccine-related paralytic poliomyelitis."

That leaves the inactivated vaccine as the alternative, but the use of the inactivated vaccine throughout the world would be no more practical or possible than it was in 1960 before the oral vaccine was approved. Cochi reflects on the likely outcome of such a choice. "There are still many countries that lack the infrastructure to deliver an injectable vaccine and most are unlikely to develop that capability within the timeframe. One has to balance the desire to change the world to inactivated vaccine against the feasibility. Not to mention the manufacturing issues—it's not clear that vaccine producers would want to increase their production capacity for a vaccine with an uncertain future. And of course, there is the cost of the inactivated vaccine versus the oral. It's likely to be prohibitive if one were to consider universal coverage, even if the infrastructure existed."

The expense associated with the decision to continue immunization would be substantial, regardless of the vaccine chosen. The population of people susceptible to polio continually renews itself—by about 130 million children each year. The current global polio immunization program costs $US 1.5 billion annually; the officials who set immunization policy have to wonder where that money will come from beyond 2005. After all, polio eradication was sold to the world in part like smallpox eradication, as a way to free up scarce health care resources.

The answer to both the risk and the expense of vaccination is, of course, to stop vaccinating. But this will be much harder to do than it might seem. It will be hard, perhaps impossible, to know when the poliovirus is gone for good, and as was the case with smallpox, many countries may be unwilling to stop vaccination. Once routine vaccination stops, vulnerability begins to grow. As long as the poliovirus is kept out of the world, this doesn't matter much. But if the virus were to be reintroduced, either accidentally or intentionally, epidemics of poliomyelitis could be expected to leave at least 0.5% of these children paralyzed. Assuming the worst, that the virus infects all 130 million children in one year, there could be 650,000 paralyzed children. With stakes as high as these, no one wants to be wrong.

Unfortunately, the possibility is real that the virus could get back into circulation. The most likely means would be the accidental release of the virus from a laboratory. Repositories of the wild-type virus exist literally around the world, in diagnostic, research, environmental, and teaching laboratories. This profusion makes life hard enough for those public health officials who monitor and control the poliovirus, but to make it harder still, many of the labs that are storing the virus today don't know it. The virus is frozen in stool specimens, water samples, and other materials collected for reasons entirely unrelated to polio, such as research studies directed at other intestinal agents. No one checked the samples for poliovirus because no one was looking for it.

Vaccine production poses a similar threat of accidental release because both the killed and the attenuated vaccine are made from the poliovirus. If production of either type continued, there would always be the risk of a laboratory worker becoming infected and reintroducing the virus into the community, an event that has already occurred several times.

By comparison, containing the smallpox virus after eradication was fairly easy. Because variola is such a severe biohazard, by the time transmission was broken in the late 1970s the virus had already been restricted to a few laboratories. Although some of these resisted giving up their stocks of the virus, a tragic laboratory accident in Birmingham, England, in 1978 convinced the holdouts to turn their stocks over to WHO or destroy them. A second difference also made containment easier: the smallpox vaccine is not derived from the smallpox virus but from a closely related virus that poses almost no threat to humans.

A less likely but perhaps more frightening scenario involves the intentional release of poliovirus—bioterrorism, in other words. The experience with the smallpox virus also puts this threat into perspective. In 1992, Ken Alibeck, a Soviet scientist who defected to the United States, revealed to the world that the Soviets had continued to work with the variola virus, finding ways to weaponize it and producing multiton quantities, in spite of a bioweapons treaty to which they were signatory. Many officials now believe that stocks of the smallpox virus exist in as many as 12 other countries with known bioweapons programs. In a world where conflict has become increasingly "dirty"—that is, aimed at terrorizing civilians—the threat of biological weapons is no longer hypothetical. There is no assurance that, after immunization stopped, the poliovirus would not become a weapon as well.

A third reservoir of the poliovirus is in a small but significant number of immunocompromised people. In people who have normal immune systems, the virus is excreted for only about three to four weeks after infection. This is not the case, however, for individuals suffering from certain genetic immunodeficiency syndromes, who may continue to excrete the virus for up to ten years. These syndromes affect as many as one out of every 10,000 people. While there is no reason to think that all these harbor the poliovirus, it is almost certain that some of them do. These hidden reservoirs could help set off disastrous epidemics once routine vaccination stopped.

Even in the face of such formidable challenges, however, plans for the end game continue. WHO has made laboratory containment a prerequisite for certification. Officials are creating an inventory of the laboratories throughout the world that might have stocks of or material containing the poliovirus. All laboratories that have such materials must implement safe handling procedures now. After the world's last case has been identified, laboratories will have one year to destroy suspect material and virus stocks, transfer strains to designated WHO laboratories, or institute high-level containment. When use of the oral vaccine is stopped, all laboratories with stocks of the virus must institute maximum level containment.

Steve Cochi, along with most of his colleagues, remains optimistic that all the challenges will be resolved. He still believes that the goal can be achieved, that the race for the last child can be won, in spite of the complexities of the final leg. "The only thing that I'm concerned about is the tendency that always exists to move on to other things, other priorities, prematurely. Continuing to keep the effort going in the face of a disappearing disease is a constant challenge. It's going to be even more of an issue in the next months and years. The sooner we finish, the better off we'll be and the more reassured everyone will be that we can go the last mile."

Another Virus, Another Vaccine: the Biologic Perspective

Another Famous Case

It is more than reasonable to suspect that Sir Walter Scott's lameness was a result of paralytic poliomyelitis. In fact, his account is among the first to clearly associate certain aspects of the disease, like the presence of fever, with the paralysis and later withering of affected limbs. His personal memories reveal much about how the disease was viewed at the time.

> I showed every sign of health and strength until I was about eighteen months old. One night, I have been told, I showed great reluctance to be caught and put to bed, and after being chased about the room, was apprehended and consigned to my dormitory with some difficulty. It was the last time I was to show much personal agility. In the morning I was discovered to be affected with the fever which often accompanies the cutting of large teeth. It held me three days. On the fourth, when they went to bathe me as usual, they discovered that I had lost the power of my right leg. My grandfather, an excellent anatomist as well as physician, . . . and many others of the most respectable of the faculty, were consulted. There appeared to be no dislocation or sprain; blisters and other topical remedies were applied in vain. . . .
>
> The impatience of a child soon inclined me to struggle with my infirmity and I began by degrees to stand, to walk, and to run. Although the limb affected was much shrunk and contracted, my general health, which was of more importance, was much strengthened by being frequently in the open air, and, in a word, I who in a city had probably been condemned to helpless and hopeless decrepitude, was now a healthy, high-spirited, and, my lameness apart, a sturdy child—*non sine diis animosus infans.*
>
> From J. G. Lockhart: *Memoirs of the Life of Sir Walter Scott, Bart.*

The Poliovirus

Compared with other viruses that infect humans, the poliovirus is elegantly simple and extremely small: a single strand of nucleic acid (its genome) surrounded by the 60 proteins of its icosahedral (20-faced) outer shell. Imagine a soccer ball only 30 nanometers wide—30 billionths of a meter—and you'll have a fairly accurate conception. The smallpox virus, in contrast, which is still far too small to be seen easily with a conventional microscope, is more than ten times as large.

The virus moves from person to person along the fecal–oral route. Polio is an enterovirus—it thrives in a person's gut—and millions of virus particles leave the body along with the other matter in the gastrointestinal tract. When crowding and poor hygiene prevail, as they have in many poor countries for centuries, then so does the virus. Infants are infected very early in such countries, and almost always harmlessly. In developed countries, even in the days before a vaccine was developed, improved hygiene and efficient sanitary systems made infection much less common. These rarer infections were also more dangerous, however, because the older a person is when infected, the more likely he is to develop paralytic polio.

The first cells susceptible to viral attack are in the throat, but the poliovirus has evolved to survive the acidic trip through the stomach in order to reach a second and preferred set of cells—those that line the intestine. From either spot, the virus begins its infectious cycle by first binding to a specialized receptor protein on the surface of the target cell, then crossing quickly inside.

Once inside, the poliovirus sheds its shell and takes over the protein-synthesizing machinery of the host cell. By having the *victim's* cells make *its* proteins, the virus begins the second step in reproducing itself. The poliovirus genome is perfectly suited to this task. A normal cell reproduces itself by making proteins. The genetic blueprint for these proteins is stored as DNA (deoxyribonucleic acid); this information must first be transcribed into short, gene-sized pieces of RNA (or ribonucleic acid), which then can be translated into proteins. The poliovirus genome *begins* as RNA. Once inside the host cell, almost the entire RNA genome of the virus is immediately translated into a single large polyprotein, which is subsequently clipped into a series of smaller proteins, each with a particular function.

As one protein prevents the host cell from doing its normal business, another protein begins to make copies of the RNA genome of the virus. A group of proteins forms a shell around a copy, and a new virus particle is complete. In ten hours, a single cell can produce as many as 1,000 new particles; the host cell dies and these particles are released, either to infect new cells and create more particles or to leave the body—in saliva or, more commonly, feces.

The Invisible Disease

With the poliovirus, it's important to distinguish between infection and disease, because in only a small fraction of the population does one lead to the other. Even in the developed world, where paralytic polio was more common per capita, the number of people stricken was, in absolute terms, fairly small—which, of course, did absolutely nothing to lessen the terror people felt during the epidemic years. On the contrary, the random nature of the disease probably only made things worse by encouraging the belief that something could have been done to avoid it. Polio seemed so capricious and cruel as well as horrible. Four children are in the same family, living in the same house, eating the same food, protected by the same parents using the same precautions—yet only one falls ill. What could it mean?

In the unlucky few, the virus moves from the gut to the cells of the central nervous system that power muscles. The official name of the disease—too bulky for everyday use—is poliomyelitis, for the swelling it causes in the gray matter (the *polios*) of the spinal cord. A victim might at first have a headache, a stiff neck, and a sore back. After these flulike symptoms comes paralysis, as the infection blocks the normal function of the nerve cells.

The damage to these cells is sometimes temporary, but unused muscle will begin to atrophy. If the case lasts long enough, the affected muscles may never recover fully.

Often the paralysis first affects the legs, and stories abound of people who discovered their polio when they tried to get out of bed one morning and found that they couldn't stand. Sometimes, however, the paralysis spreads to the nerve cells that operate the muscles in the throat and soft palate, paralyzing the victim's ability to talk and swallow. In the most devastating form of the disease, bulbar poliomyelitis, the nerves that operate the diaphragm and other respiratory muscles are affected, leaving the victim unable to breathe. It was this last group for whom the iron lung was so important—a giant respirator that enclosed the patient from the neck down and forced air in and out of the lungs.

The long-term effects of the disease are as unpredictable as its onset. Some victims, after weeks in the respirator and weeks of recovery, walk away with no outward signs of their ordeal. Some walk, but only with assistance, and some never walk again. In addition to those confined to wheelchairs, a surprising number of people have spent their entire lives in respirators. And finally, 25 to 30 years after recovery, some people suffer from advancing muscle fatigue and pain, a condition now recognized as post-polio syndrome.

The Incredible Mutable Virus

RNA viruses are notorious for their ability to mutate. Consider HIV. One of the challenges in developing an effective vaccine against it is that the virus changes the structure of the outer proteins that our immune system "sees" with such regularity that any vaccine candidate protein loses its effectiveness very quickly. The poliovirus is no exception in terms of its ability to mutate, but ironically, the virus seems to be biologically incapable of surviving mutations that affect the outer proteins against which we develop immunity.

The poliovirus exists in only three types, referred to as serotypes 1, 2, and 3. The three kinds differ in their outer proteins, so immunity to one doesn't protect against the other two. All three types of poliovirus are represented in both the oral and killed vaccines that we have used since their development in the 1950s; their outer proteins have not changed as a result of a mutational event in over 40 years. Compare this with the influenza virus, for example, where mutations that alter its outer protein shell necessitate our creating a new vaccine against the circulating strains on a regular basis. This suggests that something important biologically about these proteins in the poliovirus demands they remain unchanged—a factor that has contributed to our ability to immunize the world's population of children with the same vaccine that was developed decades before.

The Future for Global Disease Eradication

The bottom line is that eradication attacks inequities and provides the ultimate in social justice.

WILLIAM FOEGE (2001)

A leader is best when people barely know he exists. . . . When his work is done, his aim fulfilled, they will say, "We did this ourselves."

TAO TE CHING

Measles, whooping cough, cholera, and scores of other infectious diseases take thousands of lives every day. This is particularly true in the poorest countries, where the burden of disease imposes huge costs and slows economic development. Programs to control these diseases are a mainstay of international public health efforts and will continue to be well into the foreseeable future. Eradication programs, on the other hand, are more recent and more controversial. While eradicating a disease is the ultimate achievement in disease control, the move from control to eradication requires an extraordinary commitment, a commitment that cannot be made lightly.

Two different worldviews fuel the controversy. The first sees eradication as the ultimate in disease control, a strategy that offers an opportunity to improve the well-being of everyone, at least in some small and just way. To those members of the public health community, the rewards of eradication are primarily humanitarian. The choice to move from control to eradication should be made whenever eradication becomes feasible—"If we can do it, we should do it." The second view sees an eradication program as a high-risk strategy that inevitably diverts resources and attention away from broader problems. Many people today live without access to trained health professionals or to basic health services, without adequate nutrition, without clean water, and without certain basic human rights. From the second point of view, the resources needed to convert a control program into an eradication program would be better spent on controlling the diseases with the greatest impact, such as HIV, tuberculosis, and malaria, or on tackling some of the even broader issues—"we should insure that we obtain the greatest possible improvement in overall public health for each dollar that we spend."

Even for the second group, arguments support eradication as a public health strategy. Eradication, for example, may be financially attractive. After eradication, the ongoing costs of controlling the disease as well as the costs from caring for the sick go away, and years of productive life are regained for those who would have been afflicted. An eradication campaign has the proven ability to mobilize resources that would not otherwise be available for improving global health, and the cry to defeat an infectious enemy that is killing our children can engender the political and social resolve necessary for victory. A similar cry for clean water, for now at least, is less powerful. An eradication strategy offers a finite enterprise with a clear endpoint, an undertaking with measurable goals and a relatively quick conclusion. Eradication programs, as challenging as they are, appear much more manageable in contrast to ongoing control programs or efforts to improve the health care infrastructure.

Eradication programs do have a significant downside. They are riskier, for example, than control programs. Control programs are ongoing, and program directors can change tactics as infectious agents develop resistance or as at-risk population behaviors change. While the resource requirement is ongoing, all is not lost even if tactics must be changed. Eradication programs, in contrast, are more vulnerable to outright failure. An eradication program that fails to meet its objectives may simply cease to exist. Even if failure means returning to a control program, many of the resources used in the attempt to achieve total eradication are lost. The resources used in the attempt to eradicate malaria, for example, contributed little to the health of malaria victims after the program was terminated. Those resources might have had a far more lasting impact had they been used to build health care infrastructure.

The complexities of these two views go well beyond these issues, and the world public health community continues to debate heatedly the appropriate allocation of resources between the two strategies. Ultimately, however, decisions are made in the real world of global politics, social and cultural pressures, and financial realities. Within this context, eradication programs will remain controversial, but they will also almost certainly continue to play a role in global health strategies. For one thing, public health has bred a generation of men and women who know that the impossible is sometimes possible—that they can pull a disease out by its roots and banish it, however tenuously, into the history books. They have experienced the social good that can come from an eradication program, and they will continue to advocate for new ones.

Borrowing from Winston Churchill, eradication as a global health policy—much like democracy as a form of government—may be the worst policy except for all the rest. When eradication is achievable, it may be the most politically expedient mechanism for generally improving the health and well-being of the world's population. On these grounds, it becomes critical that public health leaders have sound ways of assessing whether a disease can be eradicated so as to ensure, as much as possible, that an eradication program will be a success.

This book has focused on three initiatives to illustrate the biologic, financial, political, and social complexities of disease eradication—malaria, smallpox, and poliomyelitis. As we examine the future of disease eradication in this final chapter, we see that the lessons learned from these and similar efforts will help determine the likelihood of success of coming campaigns. These lessons focus on four interrelated strategic issues; how well any given eradication strategy addresses them will decide what those coming campaigns are. The four issues encompass the following:

- Biologic feasibility—How robust are the strategies to stop transmission of the agent and to put a post-eradication plan in place?
- Financial resources—Are sufficient monies available from national and external funding sources, and are external funds distributed equitably?
- Political will—Do world leaders and decision makers view the disease as a serious health threat?
- Social benefit—Does the eradication strategy support broader health goals, including building and sustaining an equitable health care infrastructure?

Biologic Feasibility

Before the international resolutions, the billion-dollar budgets, and the armies of volunteers, eradication must begin with biology—with the pathogen and an effective way to stop it. Everything else rests on this. Unfortunately, however, this most basic part of the program is by no means the simplest. In fact, finding a feasible way to attack or frustrate a pathogen is so difficult that most diseases, as of this writing, are far beyond our power to eradicate. At least three important criteria predict biologic feasibility: a method for interrupting transmission, such as a vaccine; a means of detecting levels of infection that can lead to disease transmission; and an ecologic niche in which people are essential to the pathogen's survival.

Once these three conditions have been satisfied, it then becomes necessary to develop a strategy of attack. Biology is central here as well, and unique features of each disease and its agent preclude any formulaic approach. Even when all three criteria appear to be satisfied, a strategy based on them may not work as anticipated in the field. The highly effective insecticide DDT and the antimalarial drug chloroquine, for example, seemed to offer perfectly viable methods for interrupting transmission of the malaria parasite. Although the parasite relies on the mosquito for part of its life cycle, people still are critical to its survival. The resulting strategy developed for the eradication program was to reduce the mosquito population with the insecticide, thereby decreasing spread of the disease, and then find and treat the remaining infections with the antimalarial drug.

The failure to appreciate the biologic nuances of parasite and mosquito in their most formidable ecologic setting—the impoverished tropical and subtropical regions of the world—made the spray-and-treat method a well-intentioned but unfortunately wrong strategy. Over time, this strategy accomplished less and less. Targeting the vector had worked well in a variety of settings where land management, housing, and general access to health care had been improving long before or in concert with the eradication effort. Unfortunately, targeting the mosquitoes had never been tested in the most challenging arenas. Virtually every man, woman, and child in certain parts of sub-Saharan Africa, for example, is infected with the parasite, and the mosquitoes proliferate there with an ease not seen outside the tropics.

The smallpox eradication program grew out of a highly successful program in the developed world. The smallpox vaccine provided the eradicators with an effective method of stopping transmission, and the virus's ability to reproduce outside people is extremely limited. So once a sufficient number of people had been vaccinated, the virus had nowhere to go. It simply burned itself out and disappeared. The smallpox campaign proved that a means for detecting infection—a reliable method for surveillance—was an important biologic feature. For smallpox, this was relatively easy: tests were usually not needed; a visual inspection was all that was necessary. Knowing quickly and accurately who was infected allowed the eradication program to focus its efforts, particularly during the final phases; equally important, surveillance made it possible to measure success or failure.

Even with all the biologic criteria seemingly met, the smallpox eradication program could have fallen victim to the same failure as malaria had its champions pursued the initial strategy developed in the industrialized countries—immunizing 100% of the endemic world. The commitment to innovation in the field and the ability to integrate learned biologic lessons into the program probably saved the effort from failure. The most important discoveries—that the virus was not as infectious as first thought and that it could be tracked, surrounded with a firewall of immunity, and then stopped—became the campaign's central and ultimately successful strategy.

As the smallpox eradication campaign neared its end, one tricky issue remained. Although most scientists believed that smallpox was a specifically human disease, they also knew that species closely related to us, such as chimpanzees and orangutans, could occasionally become infected. No one was entirely certain that the virus wasn't lurking somewhere in the wild. Such a reservoir, should it exist, would represent a serious threat because the virus could be easily reintroduced. To confuse matters further, some poxviruses cause diseases that look like smallpox in other animals. Monkey pox, for example, occurs in west and central Africa, where the final campaigns were under way. Monkey pox is a disease specific to monkeys, but it can be transmitted to people if they come in close contact with an infected animal. Consequently, field teams made every effort to track the origins of each of the final cases of smallpox, and any hint of a disease resembling smallpox in primates was investigated. Laboratory identification of each virus isolated from people or primates became the centerpiece of this final strategy.

The polio eradication effort grew out of a similar history. Vaccines were known to be effective in stopping transmission, and with the development of the oral vaccine, immunizing large populations became simple and inexpensive. The eradication effort did take advantage of extensive field experience in challenging conditions before the global program began. First João Baptista Risi in Brazil and then Ciro de-Quadros in the Americas demonstrated that the combination of routine vaccination, national immunization days, and strong surveillance could eliminate the virus, even under extremely challenging circumstances. Without that evidence, in fact, it's not likely that the WHO would have agreed to another eradication campaign.

Polio has presented a more complex challenge in terms of the methods necessary for tracking the disease. The virus can be grown in the laboratory from stool specimens of anyone infected, but this, by itself, is not a realistic method for polio surveillance. The demands of the test are considerable, including everything from specimen collection and preservation to relatively sophisticated and expensive laboratory technology. Because of the number of silent infections, laboratories would have to culture stool specimens taken at frequent intervals from a very large sample of the population. This may be technically possible, but it isn't economically viable. Investigators ultimately based surveillance on disease features—in this case, acute flaccid paralysis—coupled with laboratory verification. Because other agents cause a disease that is largely indistinguishable from poliomyelitis, testing everyone with appropriate symptoms for the virus became the key to effective surveillance.

From the biologic perspective, however, the most challenging issue of an eradication campaign may be what to do after the campaign succeeds. After certifying that the world was free of smallpox, the WHO recommended ending vaccination worldwide and consolidating and then destroying the remaining virus stocks. The former was accomplished, although some countries continued to vaccinate for several years after the world was certified smallpox free. Many countries maintained stockpiles of the vaccine, as did the WHO, but over time even stockpiles of the vaccine were discarded. In retrospect, this turned out to be a mistake as it left the world unnecessarily vulnerable to the use of smallpox as a biological weapon. With the recent escalation of terrorist activity around the world, the international public health community is rapidly moving to replenish stockpiles of the vaccine.

Consolidation of the smallpox stocks was also achieved, but not without a certain amount of resistance as well. Fortunately, only a few laboratories were working with the virus because of the risk and the difficulties involved in growing it. A 1978 laboratory accident in Birmingham, England, in which a photographer died from smallpox, helped to convince the world's legitimate scientists to give up their remaining stocks. Even then, there was no way to assure that all the stocks had been destroyed, nor were people naïve enough to think there would ever be absolute certainty about this. In fact, the virus did turn up in several legitimate laboratories later on and was quietly destroyed.

The destruction of the official stocks of the virus remains a sticking point. Heightened concerns about use of the virus as a biological weapon and a lack of consensus among member countries caused the WHO to delay the date for destruction of the official stocks held in the U.S. and Russia for the fifth time in 2002, with no new date set. The additional time will be used to continue research and development of potential antiviral drugs, a better vaccine, and an animal model of the human disease.

The end game strategy is an even more complicated part of the polio campaign. Some of the same strategies apply: stopping routine vaccination and consolidating the virus stocks. The difficulties inherent in each of these are such that no one is discussing destruction of the virus at this point.

Financial Resources

The second aspect of a successful campaign concerns resources. While adequate funding can't guarantee success, the lack of resources means almost certain failure. Consequently, a strategy for funding an eradication program has become an increasingly important consideration. To what degree are the nations of the world, the developed ones in particular, willing to pay for a program? Are there nongovernment sources of funding? Can the funding be sustained until the end?

Both the national expenditures and the external financial resources necessary to support an eradication effort are substantial. The malaria initiative consumed between $US 2.4 billion and $US 2.6 billion (not adjusted for inflation) during its 18-year history, officially ending in 1969 and finally limping to a close in 1973. Smallpox eradication cost approximately $US 300 million, a bargain by comparison. If completed when projected, polio eradication will have cost approximately $US 2.45 billion in external funding alone; the total, which will include the money spent by each of the endemic countries, will be far higher. Each year the program goes beyond the 2005 target for certification will require another $US 100 million in external funding, plus ongoing country commitments.

The funding sources for global campaigns have changed dramatically over the history of eradication. Fifty-five years ago, virtually all the money spent for the Malaria Eradication Program (MEP) came from WHO, from bilateral agreements, and from the countries where the program operated. In fact, WHO seems not to have given much consideration to the financial pros and cons. There was general interest in malaria because, though primarily a problem of the developing world, it remained in some parts of the industrialized world as well. Eradicating a disease was a seductive idea on its own, one that inspired the imagination of many in the public health sector. Perhaps most important, however, was the fact that mosquitoes resistant to DDT were already beginning to appear. Under pressure that the only viable weapon might soon be lost, WHO launched the MEP.

Once it had launched the MEP, WHO put large sums of money from its own budgets into the program's coffers. In fact, at various points during the campaign, the WHO funds devoted to malaria eradication represented over 10% of the organization's regular budget and almost 35% of all other funds placed at its disposal. Given this enormous commitment and the disappointment it bought, it is easy to understand the reluctance of WHO to be the exclusive funding source for future campaigns.

This reluctance helped to create a number of funding problems for the Smallpox Eradication Programme. Although smallpox eradication was a WHO project, the organization's support was never lavish. One additional difficulty in funding a smallpox eradication program was tied to the disparate impact of the disease on the remaining endemic countries. If eradication is to succeed, it is essential that every endemic country mount a program, whether or not the targeted disease is high on its agenda, whether or not it stands to reap a direct financial benefit. It's understandably difficult to gain a commitment of resources from countries whose priorities may not match those of the WHO, and such was the case for smallpox. Variola minor, the mild and most prevalent form of the disease, often ranked well behind malaria, diarrheal and respiratory diseases, and malnutrition on the list of a country's most urgent health problems.

In contrast to malaria, money was a large part of the rationale for the Smallpox Eradication Programme. As the number of cases of naturally transmitted disease in industrialized countries neared zero, the developed world was spending huge sums each year, directly and indirectly, to protect itself against a disease that no longer occurred inside its borders. The argument was thus quite straightforward. The campaign itself would cost $US 300 million. It was costing the industrialized world around $US 1 billion annually to vaccinate and to maintain quarantines. If eradication could be achieved and vaccination and quarantine measures stopped, the cost of the campaign could be quickly repaid out of savings from prevention alone. The U.S., as an example, stood to recoup its outlay for smallpox eradication about once every two to three months.

The smallpox eradication program did establish another important precedent. Although most of the funding came from the WHO, the program also received contributions from the private sector and from a small number of nongovernment organizations (NGOs). Even though the contributions represented only a small part of the total budget, this was an important development, one that would become even more significant in future campaigns.

As in the case of smallpox, enlightened self-interest figured prominently in the decision to proceed with the polio campaign. Once again, the countries that will see most of the financial benefit from eradication will be the richest ones, although polio takes its toll of suffering and death almost exclusively in the poorest. The persuasive case for funding the polio campaign was based in part on the premise that industrialized countries, the U.S. chief among them, would save a considerable amount of money by eradicating the disease. Current estimates place the potential global savings at $US 1.8 billion annually, a large part of which will be derived from stopping vaccination worldwide.

In terms of bringing new sources of funding into a global health program, the polio eradication initiative has, in fact, set the standard. When Rotary International accepted the campaign as its own, the organization and its members pledged unprecedented levels of support. By the end of the campaign, Rotary will have contributed close to $US 500 million, plus countless volunteer hours in every endemic country in the world.

Rotary's commitment was also one of the factors that convinced a reluctant WHO leadership to go forward; in this respect, the polio campaign is unique because an NGO led the way. More recently, foundation partners such as The Bill and Melinda Gates Foundation and the United Nations Foundation, established by Ted Turner, have made substantial contributions as well. The list of additional NGOs and corporations that have contributed funds, volunteers, transport, vaccine, and help with surveillance is lengthy.

The disparity between the rewards of polio eradication is immediately apparent: piles of money for the industrialized world, freedom from a serious though relatively rare disease for the nonindustrialized world. Country contributions reflect this disparity. Although early in the campaign most countries in the Americas paid for upwards of 80% of their eradication costs, this has not been the case in the poorest countries in Asia and Africa. In exchange for their participation in the eradication program, most of these nations have received heavy subsidies and so often contribute less than 10% of the direct financial costs of the program. In most of these countries, the cost of the polio eradication program is as little as $US 0.025 to $US 0.05 per capita and represents 1% or less of their total expenditures on health.

That the bulk of the costs of an eradication program should be borne by the countries with the most to gain—in this case, the industrialized world—seems only fair. This has been a persuasive argument when applied to polio, and it will very likely figure prominently in future discussions of eradication proposals.

The only other WHO ongoing eradication program, the eradication of guinea worm, offers a different example of the public–private partnerships that have become key to global health initiatives. Guinea worm disease, or dracunculiasis, is caused by a parasite, a worm that can reach a meter in length before it emerges painfully through the skin of its victim. The parasite spends part of its life in a crustacean that looks like a tiny flea. People become a part of the parasite's life cycle when they drink water containing the infected "flea." Guinea worm disease can be prevented through a simple measure—drinking clean water. Because humans are an essential part of the parasite's life cycle, preventing humans from getting the disease would effectively eradicate the worm.

In 1987, the Carter Center, founded by former President Jimmy Carter and his wife Rosalynn in Atlanta, Georgia, became an advocate for the idea that guinea worm should be eradicated. The guinea worm initiative fit well within the center's mission to reduce conflict and to alleviate suffering in the world, and under the leadership of Donald Hopkins, a seasoned smallpox warrior, as the Center's associate executive director, a concerted effort began to mobilize support. WHO asked the WHA to approve the eradication program at its May 1991 meeting.

At first, the guinea worm eradication program focused on improving the rural water supply in the affected countries, and it operated out of funding from the WHO's Water and Sanitation Decade. With experience, however, the field staff found a faster and less expensive intervention—it provided the people living in the endemic areas with simple water filters. An unanticipated partner helped to secure the resources needed—E. I. du Pont de Nemours & Company. The company has supplied the special nylon used to make the filters, a fabric that has had no other use, for over a decade. A section of the company—Precision Fabrics Group—even produces the filters. Other corporate citizens and most recently the Gates Foundation have also contributed resources, but without the early commitment of the Carter Center and DuPont, the campaign wouldn't be nearing its successful end.

DuPont's actions illustrate the growing spirit of corporate citizenship that is driving new resources into global health initiatives, particularly those that lack the direct economic payoffs of smallpox and polio. To some degree, Roll Back Malaria will be a test of this new altruism, but so will other programs under way. Merck & Co, Inc. has donated many millions of dollars' worth of drugs to treat river blindness; the company recently expanded its contribution to cover people suffering from lymphatic filariasis. Both river blindness and lymphatic filariasis are diseases that affect predominantly people living in poverty in tropical and subtropical regions of the world. GlaxoSmithKline has initiated a pilot program to treat trachoma, and Novartis has been instrumental in forming an alliance to tackle the last cases of leprosy. Both of the latter are also donating the expensive drugs needed to control these diseases. None of these programs has yet reached the point where eradication is feasible, but one or more of them may within the next decade.

Corporate altruism does present certain challenges for the future. If philanthropic interests don't match the major goals of the world health community, what is the impact on existing programs? If the WHO is devoting its resources to controlling malaria, for example, and a corporate donor wants to fund a large program to control trachoma, how many of WHO's resources might be diverted from the former program? Can the different interests of the donors be reconciled or even coordinated so as to diffuse this negative impact? At what point do corporate initiatives, however well intentioned, become so disruptive of existing programs that they have to be discouraged? How long can corporate philanthropy be sustained if the goal is control instead of eradication? Currently, a nonprofit organization called The Task Force for Child Survival and Development is attempting to coordinate the various donor programs on WHO's behalf, but reconciling the varied interests of corporate and institutional donors is already presenting an expected challenge.

Another trend is emerging that will almost certainly influence the funding strategies of future programs. Roll Back Malaria perhaps best exemplifies the trend. All the countries where malaria transmission remains most intensive are among the poorest in terms of gross domestic product (GDP)—in other words, countries that are the least able to pay for such a campaign themselves. Consequently, there has been little interest beyond the national contributions already going to the WHO in supporting control programs in the malaria-endemic countries.

Given this climate, much of the rationale used to encourage funding for RBM is based upon malaria's probable impact on economic development. With minor exceptions, malaria and poverty go hand in hand, but there was no evidence to suggest that malaria itself causes poverty. Economists John Gallup and Jeffrey Sachs, working at the Center for International Development at Harvard University, maintain that areas where malaria has remained endemic have per capita GDP growth rates

1.3% lower than do comparable regions that have essentially eliminated the disease. This discrepancy remains even after such factors as initial income and overall health are taken into account. To put that into perspective, if malaria control had been achieved in Africa 35 years ago, the continent's current GDP of $US 30 billion might easily have increased, through incremental growth, by as much as 54%, or $US 16 billion. That amount is at least five times greater than the total amount of annual development aid received by African countries in recent years.

The promise of these potential markets was enough to attract the attention of the G8, the eight largest economies in the world. At their May 1998 Summit, G8 representatives pledged support for the renewed malaria efforts. Pledge or no pledge, however, substantial funding gaps remain for the countries farthest along in their planning; the promised resources have not materialized. Each country is responsible not only for developing its own strategic plan, but also for identifying funding sources. The 12 African countries that developed the earliest strategic plans have not gotten the necessary funding commitments, reporting only $US 51 million out of budgets that total almost $US 700 million. At this point, it is not clear how, or whether, this gap will be closed.

Given this array of government and nongovernment funding sources and competing national, international, and corporate interests, it's safe to say that amassing sufficient funds in the future will be a complex undertaking.

Political Will

The success of any eradication program depends heavily on a third factor—political will. An eradication campaign is all consuming; every country in the world must be prepared and willing to commit to the initiative, including providing some of its own resources. Further, today's eradication initiatives depend upon multiple government sectors, not only the health ministry but also the transportation and communications sectors, and even the military. This means that heads of state must be convinced that the campaign will be worth the expense, not to mention the headaches.

Obtaining global political commitment for disease eradication is a tricky business. It would be virtually impossible to wage an eradication campaign without the endorsement of the WHO, so gaining the organization's support becomes a critical first step. Further, a program is going nowhere within WHO until the director-general can be convinced to recommend an eradication program to the World Health Assembly. This in itself is no easy undertaking. Over the past five decades, only five eradication programs have gained approval—malaria, yaws, smallpox, polio, and guinea worm—and only the malaria and yaws programs had the unconditional support of the director-general when they began.

The high-profile malaria program and the lesser-known yaws initiative left their mark on the WHO in this regard, or more precisely, on directors-general. The failure to eradicate either disease was extremely costly in terms of credibility. No director-general since has wanted to leave the organization similarly vulnerable. Nonetheless, the combination of powerful advocates—the Soviet Viktor Zhadanov for smallpox, for example, or Jim Grant and Rotary International for polio—and demonstrated success in the field have been sufficient to move the politically conservative leadership of the organization forward. As one prominent leader in the polio eradication initiative puts it, "They are primarily a political organization, and the political alignment has to take place to push them safely in the right direction."

In theory, at least, once the director-general gains the approval of a formal resolution by the delegates of the World Health Assembly, every member country is committed to an eradication program. The reality is somewhat different. As D. A. Henderson describes it, "Many people are happy about voting for these programs, but the Ministers of Health often do not have the support of their countries. WHA approval thus does not constitute buy-in from all the countries represented. Many times there has been no consideration by high-level officials within a country at all."

Gaining commitment ultimately is a country-by-country effort; "all politics are local" becomes the operative phrase. Competition for resources within a national budget can be fierce. Resistance can come from a variety of quarters, including the field staff of the WHO itself. Unstable governments may dissolve. Countries in conflict, particularly where no central governing authority exists, make access anything from challenging to impossible. Within this realm, the role of strong advocates at many levels of society cannot be underestimated.

The requirement for national political support was appreciated from the outset of disease eradication. Hence, the head of every National Malaria Programme reported directly to the head of state. While this ensured some level of national commitment for the program, it also created its own difficulties by separating the malaria teams from the public health services. Thus it was a pattern that wouldn't be repeated in the ensuing programs. Other mechanisms had to be found.

Each eradication campaign since has had its own unique challenges in gaining and maintaining the necessary country-level support. Ethiopia, for example, was the last country to commit to the smallpox eradication initiative. The minister of health, acting on the counsel of the USAID malaria advisors, actively resisted the program. Only through luck and happenstance were WHO program directors able to enter the country and then get a copy of their proposal into the hands of Haile Selassie, the country's less-than-benign emperor. Selassie, to his credit, forced the program through. Without his intervention, the entire smallpox eradication program might have failed.

Fortunately, individuals who can make a difference have always come forward. Sometimes they are world leaders, such as former U.S. President Jimmy Carter. Nigeria is among the countries where guinea worm disease is endemic, and it represents a special challenge because of its size and population density. Although some were already at work to convince the Nigerian government that guinea worm was an important cause of disability in the country, it was the involvement of the Carter Center, and of Carter personally, that gained the ear and ultimately the commitment of Nigeria's President Ibrahim Babangida to the program.

The polio eradication program has had its own successes in this regard. In 1993, President Jiang Zemin administered the first dose of oral polio vaccine during a nationally televised ceremony in the first national immunization day in China. His presence signaled support for the eradication effort at the very highest level in China and brought the most populous country into the program. Similarly, heads of state have played a crucial role in sub-Saharan Africa. With Africa lagging behind and with many countries left uncommitted to the WHO strategy, the regional director for the African Region established the Committee for a Poliomyelitis-free Africa. It comprised many prominent figures, including then South African President Nelson Mandela, who agreed to chair it. Shortly thereafter, the African heads of state attending the summit of the Organization of African Unity approved a declaration in support of eradication, which went a long way toward bringing resistant countries on board.

Other individuals who lack the political cachet of a former U.S. president or a sitting head of state have found their own way to influence governments. The former French colony of Côte d'Ivoire on the west coast of Africa represents a case in point. In December 1999, Côte d'Ivoire experienced the first military coup in its history. The government of then President Henri Konan Bedie was overthrown, and the leader of the junta, General Robert Guei, assumed the presidency. The polio eradication initiative was well under way, with an important national immunization day (NID) scheduled for January 2000. In the wake of the coup, however, no one, not even the minister of health, knew whether the NID would take place, or even what the status of the larger initiative would be. A young and articulate spokeswoman from Rotary International managed to gain access to the new president's wife. She described working in a village during a previous NID in words that appealed to Mrs. Guei as the mother of her own young children. "I saw a woman disabled by polio walking toward me. She had two children by the hand and she carried one on her back. She wanted to make sure that her children didn't suffer like she had; she was struggling to keep her children safe from the disease that had crippled her." She explained that the upcoming NID was an important part of that struggle.

The young volunteer was able to move Mrs. Guei with her story, and she took the request for continued commitment to her husband the same day. The NID went forward as scheduled, with the president personally kicking off the campaign. Côte d'Ivoire has been free of polio since 2000.

The popularity of the polio campaign in particular seems to have increased the political value of health care; many national leaders have realized that it's good politics to support public health, eradication included. Whether this wave of interest will translate into substantially greater resources for public health, and whether the interest will be sustained for future programs, remain to be seen.

Social Benefit

Even though incontestable humanitarian benefits are derived from eradicating a disease, future programs will almost certainly be expected to do more, to provide additional social benefits. The fact that eradication programs divert resources from other public health programs has been a point of contention in policy debates for the past two decades. The tension between the two camps—those who promote targeted programs and those who support programs with broader social goals—has forced policy makers to consider more carefully the inherent tradeoffs in each approach.

The Malaria Eradication Programme set the stage for this debate. The strategy its leaders adopted did nothing to develop sustainable infrastructure or improve health care capacity. If anything, it did the opposite, diverting enormous resources, both financial and human, away from many other WHO and national initiatives ongoing at the time, including smallpox eradication. By design, the MEP existed completely outside the health care system of the endemic countries. This worked to the advantage of the Malaria Eradication Programme, allowing it to concentrate its efforts and guard its resources—in some cases, as in Ethiopia, with an almost religious fervor. The separation also, however, foreclosed any opportunity to bolster existing health services. The program left virtually nothing behind in terms of increased capacity. The spray teams and those who managed them gained little or no experience that would prepare them for other health-related activities.

Despite a philosophic disposition toward broader health goals, the people directing the intensified smallpox program found both their resources and their time severely limited. As a result, they were forced to be highly focused, and this left little time or money for addressing broader goals. The outcome again was that little of general use was left behind once the program was completed. The main exception to this was significant, however: the smallpox eradication program trained a group of health professionals who knew how to conduct disease surveillance and who understood its importance to sound public health action. The smallpox warriors came to believe that eradication of other diseases was achievable and that eradication should be an important part of public health strategy.

The position that international public health should concentrate on capacity building was set forth toward the end of the smallpox campaign and may have resulted from the impact of both the malaria and smallpox initiatives on more general goals. In 1978, the WHO conducted a major conference at Alma-Ata, Kazakhstan, that outlined a global agenda for developing primary health care. Those who attended left firmly committed to the broader goals defined at the conference. This commitment influenced certain members of the public health community to such a degree that even an expanded program to improve immunization practices for a variety of childhood diseases was viewed as a threat to strengthening systems for delivering primary care. Needless to say, the people most firmly entrenched in the primary care camp did not view eradication with friendly eyes.

Both contemporary eradication programs, the polio and guinea worm initiatives, reflect an effort to extend social benefits beyond the most immediate one—relief from a specific disease. In fact, the WHA specifically committed the WHO to global polio eradication in ways "which strengthen national immunization programmes and health infrastructure." Although authorities within both programs acknowledge that they have not taken advantage of every opportunity to strengthen basic health services, each program has helped in significant ways to build infrastructure at the national, regional, and local levels. The degree to which each helps to build health services or, conversely, to threaten the integrity of existing services and broader goals will figure prominently in the debates over future programs.

Social benefits may accrue in five areas: national health care policy, the organization and structure of health care systems, human and financial resources, the administration of existing health care programs, and the development and introduction of appropriate health technologies. Public health researchers have made some effort to measure the impact of the polio and guinea worm eradication programs in these regards, although this has proved challenging. Every country is different, and the effects, good and bad, are not always readily apparent. Despite the fragmentary nature of the evidence, an assessment of the successes and failures in these five areas will determine the requirements for approving future programs.

Redefining National Health Policy

A country's national health policy sets strategies and determines how resources will be allocated. In an ideal world, the most pressing health issues would drive the formulation of that policy. In reality, where health care systems are least developed, health policies are often vague and outdated if they exist at all. The pressures of an eradication program have the potential to realign national priorities and thus either improve or weaken existing policies.

In this regard, the polio eradication program has had a negative effect on national policies in some developing countries. Poliomyelitis was not one of the most significant factors in the burden of disease. Perhaps appropriately, many national polio control programs prior to the eradication campaign were given a relatively low priority. This low priority, in fact, meant that national immunization days and other supplementary activities were necessary if eradication goals were to be met. The polio eradication campaign, as a result, forced some nations with established health policies and priorities to divert resources. The staff managing immunization programs in Mozambique, for example, committed large amounts of time to planning national immunizations days, to the detriment of other programs they were supervising.

On the other hand, an eradication program can also help to realign outdated or poorly structured priorities. Guinea worm disease was not a top priority in most of the 18 endemic countries before the eradication campaign began. Because this disease usually affects people in remote villages, health ministries located in large cities were not fully aware of its impact. Where agriculture forms an important part of a nation's economy, the fact that 60% of the agriculture workers can't go to the fields because of guinea worm disease is significant. An important national realignment of resources was made in countries such as Nigeria once that impact had been defined.

Improving Health System Organization and Structure

Experience has shown that an eradication program can be integrated into an existing health system in such a way as to strengthen the management of that system. This was certainly true in some of the Central and South American countries, where the polio eradication program helped to develop stronger lines of authority and better management practices. Similarly, the guinea worm program has actually helped to create health services in some of the 18 endemic countries. Workers discovered villages in the remotest parts of some of those countries that were not even known to exist by other health workers. Because the guinea worm teams were working within existing health care organizations, these villages were integrated into the country's health care management structure.

The potential for benefit depends to some degree on the health system already in place when the eradication program begins, particularly its level of development. Where an established structure with well-defined lines of authority, responsibility, and accountability exists, as in India, an eradication program may add very little. What it can do in these circumstances is provide an opportunity to demonstrate effectiveness. In Cambodia, for example, a newly revitalized ministry of health was able to organize and implement highly successful national immunization days, increasing its public recognition. In Yemen, the organization of national immunization days was so successful that it resulted in a tenfold budget increase for preventive health services.

The challenge, of course, is that in some countries, the health system may be so weak that the demands of the eradication program necessitate a separate organization—a so-called parallel structure—just to administer the program. This may well turn out to be the case for certain of the most fragile African countries, particularly those where no central authority exists, such as Angola and Sudan. Both the polio and guinea worm initiatives have been particularly difficult in these countries, where parallel structures were the only alternative. If some part of these structures remains after eradication has been achieved, then an important opportunity will have been realized.

Walter Dowdle, a former Deputy Director of the CDC who now works with The Task Force for Child Survival and Development, discounts the notion that shifting national priorities is necessarily a bad thing. "If you went to some of these countries and saw what they have on their shelves and what they have to offer . . . even though they were diverted from what they would normally do, the diversion was from someone sitting there and explaining that no drugs had come in."

Increasing Human Resources

Successful eradication programs and effective health services all rely on people, from the head of state to the health worker in the remotest village. Eradication programs can be designed to build these human resources: a head of state can become a strong supporter of public health; a health care worker in a remote village can gain the knowledge and technology to do his or her job better.

Many advisors and WHO consultants have experienced the enthusiasm and commitment that an eradication campaign can generate. The local public health communities are often an underused resource, and an eradication program can bring new energy and a sense of purpose to people who work without tangible rewards. Steve Cochi, director of the CDC's Global Immunization Division, reflects, "The polio campaign has been very visible. It's concrete, you can measure its success. It's made people feel it's possible to come up with an ambitious target and succeed." Such motivation can help push people to higher levels.

Eradication programs can also lead people to develop new skills and knowledge, considerably expanding the pool of trained health care workers. The guinea worm program, for example, has recruited new workers and provided additional training to existing health care workers in more than 15,000 remote villages. Many of these workers have already taken on other programs, like river blindness and lymphatic filariasis, in addition to their responsibilities for guinea worm.

The polio initiative has had a similar effect. Cochi takes some pride in the human legacy he believes has resulted. "Polio eradication . . . has given rise to a field staff, especially on the WHO side, down at the country level. It's a huge human resource that has already been called upon to deal with other immunization priorities. Harnessing that field staff, which is now about 1,600 strong, and [broadening] their scope of work over the next years is a challenge, but also an exciting place to be. They didn't exist ten years ago; they didn't even exist five years ago. I think they are the key that will unlock other major achievements in global immunization."

Dowdle shares Cochi's view that human resources are one of the greatest social benefits of an eradication program. "The future leaders of public health programs are working in polio [and guinea worm] today, just like they were in smallpox in the past. There will be people out there who have been in a successful program and who now understand the concept of goals and how to achieve them, and they're going to want to do it again. It could be in child and maternal health, it could be a lot of areas. I think that is one of the greatest legacies of the programs." It will be up to future programs to maintain and extend that legacy.

Expansion of Existing Health Care Services

Health services are often limited to urban centers in developing countries, available only to those who live in or relatively near such centers. In this respect, an eradication campaign can bring an aspect of equity that sometimes isn't present in existing services—it extends access to a much wider population. Further, an eradication campaign may add other interventions and services, such as additional vaccines and more general health education, to its primary objective.

The polio and guinea worm campaigns have both made commitments to expand their services wherever feasible. The inclusion of vitamin A supplements along with polio vaccine during national immunization days offers a useful example. Vitamin A deficiency places over 140 million preschool children at much higher risk for childhood infections, many of which are fatal. Left uncorrected, severe vitamin A deficiency by itself will cause blindness. Administering a high-dose vitamin A capsule twice a year can reverse most of the effects.

Vitamin A is in most ways a perfect match for polio eradication. The target group for intervention is the same as polio—children under five. The supplement is squeezed from a capsule into the child's mouth, just like the oral vaccine, so volunteers can readily administer both vaccine and capsule, and there is only a minor incremental cost to including the vitamin. In 1998, vitamin A supplements were delivered to over 60 million children during national immunization days. Since then, vitamin A has become a standard extension of service in almost every country. Although this strategy offers neither a permanent nor a complete solution to vitamin A deficiency, the eradication program has helped to increase awareness of the problem and to improve for the short term the nutritional status of many millions of children at greatest risk.

The nearly six-month-long "guinea worm ceasefire" that was negotiated in Sudan in 1995 offers another example of short-term benefits that can be achieved by an eradication program. Because of the civil war, health services in Sudan are essentially nonexistent. Once health care workers were allowed into the region, they were prepared to deliver much more than guinea worm surveillance and education about clean water. The workers treated over 100,000 people at risk for river blindness, vaccinated over 41,000 children against measles, another 35,000 against polio, and 22,000 against tuberculosis. They distributed 35,000 doses of vitamin A and treated 9,000 children with oral rehydration packets.

Such expansions do have practical limits. In the case of polio, any intervention added to the NID strategy has to be appropriate for the age group; otherwise, the campaign risks jeopardizing its central goal. Further, huge numbers of volunteers participate in vaccine distribution during an NID. This is possible because polio vaccination is so simple; it requires no more than placing a few drops of vaccine in the mouth of each child. All other childhood vaccines, however, require injections. So while it might seem advantageous to administer other vaccines during a national immunization day, this could not be done with the same pool of untrained volunteers. Nevertheless, future campaigns can attempt to expand services whenever feasible.

Both the polio and guinea worm campaigns will likely make broader and more sustainable contributions to health services than did either the malaria or the smallpox campaigns. In the majority of countries in which the polio and guinea worm campaigns have taken place, the trained public health workers, electronic communications, and organizational structure are in place to conduct disease surveillance more generally. In the Americas, for example, 20,000 surveillance units were created during the polio eradication program. The surveillance networks, while still devoted principally to the needs of the specific eradication programs, have already been expanded to include other diseases of public health concern, such as measles, neonatal tetanus, and cholera. Coupled with the increased awareness at the highest levels of government of the value of disease surveillance as a tool for public health action, this resource has the potential to produce significant benefits well beyond the end of the current campaign.

The challenge is to maintain support over the long term for these networks. Otherwise, one of the clearest benefits of the eradication campaigns will be lost.

Developing and Inserting New Technologies

The final area in which social benefits can accrue from eradication programs is the development and insertion of new health care technologies. One of the most striking examples of this benefit is the global laboratory network created to support the polio initiative. Polio surveillance depends first on finding patients with acute flaccid paralysis and then on confirming in a laboratory that the poliovirus is the cause of the paralysis. The need to test specimens quickly and keep them cold during transport meant that regional laboratories must be available. More importantly, these laboratories must be reliable, because false results have potentially disastrous consequences.

Not enough money or time were available to build new laboratories devoted to this purpose, so existing laboratories were equipped, and personnel were trained in the relatively sophisticated procedures necessary to detect the virus in a specimen. The result was a three-tiered system of 148 enteroviral laboratories around the globe. The first and largest tier is the system of national and subnational laboratories where primary specimens are cultured and any resulting viruses are identified. The second tier is made up of regional laboratories that distinguish vaccine strains from wild virus. Specialized reference laboratories, the final tier, perform the genetic analyses that help determine transmission pathways, the origin of importations, and the characteristics of remaining virus reservoirs. In many cases, these laboratories are already supporting surveillance activities for other diseases, such as measles and rubella.

A smaller but still useful contribution from the polio eradication initiative has been the refurbishment of equipment needed to keep both the vaccine and the specimens cold. The equipment and procedures used in the polio campaign can help preserve other vaccines and drugs that require refrigeration, thus lowering one of the barriers to delivering health care in developing countries.

The guinea worm program has made a different kind of contribution in this regard. The eradication program grew out of WHO's Water and Sanitation Decade, one of the goals of which was to provide safe drinking water. Providing safe water, especially from borehole wells, was an important way to break the parasite's transmission cycle. Once the educational networks were established, hundreds of communities mobilized to improve their water supplies. In southeast Nigeria, for example, villagers have created over 400 hand-dug wells. The safer water supply has not only stopped transmission of guinea worm, but has helped control other water-borne diseases as well. That, coupled with the establishment of community-based health education, village task forces, and surveillance by village volunteers, seems to have gone a long way toward improving primary care. As Hopkins recently wrote, ". . . the [guinea worm eradication program] has done more to improve primary health care in endemic communities than many primary health care programmes. Primary health care was not developed in most of these communities before the program began."

Of course, the same caveat that applies to surveillance systems also applies to technological improvements. If resources are not made available to sustain these advances, the social benefits will disappear along with the diseases they were designed to eradicate.

The Next Campaign

With the impending eradication of polio and guinea worm disease, world health leaders are already searching for the disease that will be the next target, and the one that seems to be gaining the most political backing is another disease of childhood—measles. There are strong arguments for targeting that disease. Measles has a significant global impact: the disease is responsible for at least 10% of all deaths of children under five years old, and it is the eighth leading cause of death worldwide. WHO estimates that there are 40 million cases in the world each year, one million of which end in death, although these numbers are soft because adequate surveillance doesn't exist in many countries. Economic analyses suggest that it costs about twice as much to treat the disease and its complications as it does to vaccinate against it. Some external funding is already flowing into measles vaccination, principally as a component of the global vaccine initiative, and it may be the desire to capitalize on this momentum that makes measles the most attractive target to some world health leaders.

An aggressive campaign to "eradicate measles from the Americas" has already been undertaken by PAHO. Using a strategy similar to the one employed for polio—routine vaccination between one and four years of age, a one-time mass vaccination of one- to 14-year-olds, a mop-up campaign to increase coverage, and mass vaccination of one- to four-year-olds every four years—public health workers under the leadership of Ciro deQuadros have broken transmission of the virus in a number of countries. Cuba had its last confirmed case in June 1993, and there have been no confirmed cases in the English-speaking Caribbean since 1997.

On the other hand, there are good reasons to view a measles eradication campaign with skepticism. Even the success stories, Cuba and parts of the Caribbean, may offer less reason for optimism than at first appears: these are, after all, small islands, with small populations. Eradicating measles in larger, less easily controlled settings will be much more difficult. While a highly effective measles vaccine provides protective immunity for at least 20 years, it must be administered by injection. This precludes the use of the armies of volunteers, the chief feature of national immunization days, since specialized skills are required. Surveillance for the disease, an essential element of eradication, is not a simple matter, either. The rash associated with measles is difficult to see in dark-skinned people, and even in those with lighter skin, the rash is difficult to distinguish from rubella. Unless there is a clear epidemiological association—direct contact, in other words—with a confirmed case, a suspected case can be confirmed only by an antibody test that requires a blood sample. Either of these technical requirements—the need to inject the vaccine and to confirm suspected cases with a blood sample—would have been enough to scuttle the polio campaign. Only after scientists developed an oral vaccine and found alternative means for diagnosing the disease did the polio program become even thinkable. Both the current measles vaccine and the diagnostic test argue for the need for simpler technology before serious consideration can be given to a global campaign.

Although some other countries in the Americas have been able to reduce the number of cases to zero using the PAHO strategy, none has remained disease free. Chile, for example, became disease free in 1993 and remained so until measles was reintroduced from Brazil in 1997. The highly infectious nature of the virus, coupled with a growing reservoir of susceptible young adults, makes importation a constant threat, even in the face of vaccination rates among one- to four-year-olds that exceed 95%. The risk of reintroduction means that any global campaign would have to achieve very high levels of coverage in a very short time; given what we know about the polio and smallpox campaigns, this strategy does not appear to be feasible. Further, the strategy used in the Americas has not been field tested in Africa, where malnutrition and other predisposing factors are much more important in the epidemiology of the disease.

An interesting feature of the debate over measles eradication is a concern on the part of some—most of whom, it should be said, are not scientists or physicians—about the safety of the vaccine. The vaccine used throughout most of the industrialized world, known as MMR, combines measles, mumps, and rubella in a series of two injections—a strategy that public health specialists hold is more effective than using three separate vaccines. A small but increasing number of parents, however, believe that the combined vaccine is linked to autism and are refusing to have their children immunized, despite the fact that there is no credible scientific evidence that such a link exists. As a result, several small measles outbreaks have already been documented in England and in the U.S. Whether such concerns will make an eradication program more likely or less likely is anyone's guess. On the one hand, eradication would make it possible to stop using MMR altogether. On the other hand, it will be hard to maintain the necessary levels of immunity throughout an eradication effort.

An additional challenge lies in gaining the political support to mount a global eradication initiative. The majority of developed countries don't perceive measles as a priority, and this will have to change before a campaign can move forward. Even though the impact of measles is greater in developing countries, many other diseases are more significant, such as malaria, HIV/AIDS, and tuberculosis. As with many of the other questions surrounding a measles campaign, whether sufficient political will can be gathered also remains unclear.

Other factors may be motivating the case for measles. A program could build upon and help to sustain the infrastructure of laboratories, surveillance and field staff, and administrative expertise that has grown out of the polio eradication program, and that would be good. Yet a measles campaign following on the heels of polio eradication might distort priorities in some of these countries even further.

At present, the risk of failure seems quite high and the cost would be measured both in economic terms and in loss of credibility within the public health community. The biologic feasibility of measles eradication has not been tested in the most challenging areas of the world, and there is little political support within the international community—most importantly, among the industrialized nations. Without strong political support, the financial resources necessary to conduct an eradication campaign are unlikely to materialize.

Eradication programs for other diseases may soon become biologically feasible. These diseases include several of the vaccine-preventable childhood diseases and hepatitis B, one of the leading causes of liver cancer in the world. At this point, however, none of the candidates is any closer than measles to being eradicable.

If public health leaders were to choose their public health investments based strictly on which diseases are the deadliest or most costly, three would stand out—malaria, tuberculosis, and AIDS. Together they accounted for 5.7 million lives lost in 2001, many of them either young adults in their most productive years, or children. These diseases predominantly afflict people in developing countries. Eradicating them would not only drain an ocean of tears, it would also increase productivity and life expectancy, both of which are central to sustained economic growth and development. Unfortunately, none of these diseases is a viable candidate for eradication, but all are candidates for aggressive control programs.

Malaria has been discussed extensively in this book. Current interventions—insecticide-treated bed nets, mosquito control, rapid diagnosis and treatment of cases, and intermittent prophylaxis for pregnant women—are sufficient to bring the disease under control if these interventions can be made accessible, affordable, and socially acceptable.

Tuberculosis offers another important target. *Mycobacterium tuberculosis*, the bacterium that causes the disease, has infected two billion people. While the majority of these two billion infections are silent, some 8.8 million people develop active disease, and 1.7 million people die from tuberculosis every year. The disease can surface at any point during an infected person's life if his or her immune status deteriorates. In effect, the tuberculosis bacterium is a time bomb ticking in the lungs of a third of the world's population.

Tuberculosis is controlled in the industrialized countries by aggressive public health measures—rapid diagnosis and effective treatment for active cases, and prophylactic treatment for newly infected people before the active form of the disease can develop. Such measures prevent the spread of the tuberculosis bacterium. Although a vaccine, BCG, is available, it prevents serious disease but does not, as many vaccines do, prevent the infection altogether.

The application of similar measures on a global scale would significantly reduce the number of new cases each year, bringing tuberculosis under control worldwide. The cost of the drugs to treat new infections is relatively small—about $US 10 per case—although the cumulative cost of the drugs would exceed $US 80 million annually. Drug resistance, which emerged two decades ago, presents its own challenge. Second-line drugs are available and effective, but they are more expensive. The external funding required for the massive scaling up to achieve control in the 22 most highly endemic countries is estimated to be between $US 1 billion to 1.5 billion.

HIV/AIDS is the third disease with major global impact. Thirty-six million people were living with HIV/AIDS at the beginning of 2001, with 92% of those infected living in developing countries, predominantly sub-Saharan Africa. Unlike the AIDS epidemic in the United States and Europe, HIV/AIDS in Africa is a predominantly heterosexual disease. Fourteen million of those infected are women of childbearing age, increasing the risk of children being born with HIV. Over 21 million people, including 4.3 million children, have died since the beginning of the epidemic. The epidemic has also left more than 13 million orphans, many of whom live on the streets.

HIV infection can be drastically reduced, through the use of condoms, for example. Condoms can be supplied extremely inexpensively, but significant cultural and social issues make their introduction and use problematic. Other, costlier steps can also be taken. Antiretroviral drugs, for example, will largely prevent transmission from an infected mother to her unborn child, and clean needles decrease transmission among intravenous drug users. Screening the blood supply for contamination will significantly reduce HIV transmission as a result of transfusion. The added cost of extending these latter interventions into the developing world would require an external investment of at least $US 10 billion annually.

Even though eradication is not a serious consideration for these diseases at present, the investment in controlling them until new interventions can be found is a rational priority. Given the choice between mounting a global eradication campaign against measles and investing in global control of malaria, tuberculosis, and HIV/AIDS, the choice is clear: the former is still quite risky, and the latter is more likely to improve world health.

Actions by WHO during the first months of 2002 indicated that the emphasis on malaria, tuberculosis, and HIV/AIDS may very well increase in the months to come. Public health authorities in Geneva believe that the political and social will may be coming together to bring needed resources to the table. This belief is based on a series of high-profile endorsements, including the members of the G8, who in 2001 committed $US 1 billion toward the creation of a global fund to fight AIDS, tuberculosis, and malaria.

While all this recent interest is promising, the move from control to eradication will depend ultimately as much on scientists as on world leaders. The key to eradication remains in the laboratory, where new interventions—in particular, effective vaccines—will be developed. This is a two-sided problem. On one side, developing new interventions requires investment. Up to this point, research funding has lagged far behind even the limited funding available for deploying the old interventions. For example, in 1998, less than 1% of the world's $US 70 billion budget for research and development went toward vaccines, medicines, and new diagnostic tools for malaria, HIV/AIDS, and tuberculosis.

HIV/AIDS does represent the exception among these three diseases. AIDS research has been heavily funded primarily because of its impact on people living in the United States and Europe. Unfortunately, the wealth of new drugs and diagnostic tests that have resulted are beyond the means of most of the infected people in the world. Incentives are limited for pharmaceutical companies to develop vaccines, drugs, and diagnostic tools for malaria and tuberculosis. These diseases also occur predominantly in the poorest countries in the world, where there is little hope of a return on investment, and neither disease is viewed as a particular health threat in the industrialized countries. With each new product averaging $US 500 million to bring to market, this will not change until some other mechanism can be found to fund research and development for these diseases. A series of public/private ventures have begun to take shape, which may help address some of the funding issues and provide the needed incentives for private corporations to invest.

The other side of the problem—biologic feasibility—is even more challenging than finding adequate funding mechanisms and alternative incentives. Despite more than a decade of well-funded, intensive research, the biology of HIV has so far thwarted the efforts of some of the best scientists in the world. While not well funded, the search for a malaria vaccine has gone on for more than three decades without success. Renewed interest is spurring research into new drugs and a potential vaccine for tuberculosis, which may prove slightly easier to tackle biologically than the other two. However, the question of biologic feasibility will remain unanswered until these new interventions have reached the point where they can be tested.

This means, among other things, that we are still a very long way from considering any of these three major diseases for eradication. In the meantime, the international public health community will continue to struggle each year to cover the enormous expense of trying to control them.

It's clear, however, that given these circumstances—limited resources, unlimited need—eradication will have a future, and that it will continue to be risky and thus controversial. Both those who favor eradication programs and those who oppose them will turn to history for support, and find it. But as long as the need remains for real and cost-effective solutions to world health problems—and it appears that this need will remain for a long time—people are unlikely to give up on the dream of eradication, even if they have to wait decades before they can try to realize that dream.

Epilogue: Voices from the Eradication Campaigns

Christopher Plowe and Abdoulaye Djimde: Malaria Warriors

Chris Plowe and his Malian colleague Abdoulaye Djimde appear relaxed in their enclave at the University of Maryland Medical School in Baltimore. They have just returned from one of their many trips to Mali where Djimde is studying chloroquine resistance and Plowe is overseeing studies that will prepare their research site in Bandiagara for a future vaccine trial.

Plowe says that life in the field has gotten easier for them. "We have a couple of houses rented now. They are around the corner from each other—simple one-story houses with several bedrooms. The kitchens are these little tiny things; the cooking mostly goes on over a charcoal stove right outside the kitchen. We have dinner together on the big front porch."

Mosquitoes are a constant presence in Bandiagara, even in the dry season, but Plowe, like most of the Bandiagarans, doesn't sleep under a bed net. He simply blasts his newly air-conditioned room with locally obtained insecticides before dinner so that he can return to a relatively mosquito-free evening. "When we first started going there, it was April with temps up to 115 degrees or so. Everyone except me slept out on the porch. It was too hot to use the nets. Before that I used to stay at the nutritional medical center in little stone igloos, which soak up the heat all day long. I would drag a bed out and sleep under a tree until the noise from night creatures became so loud that I had to retreat back to the igloo."

The work schedule for the team can be grueling. While Plowe is there primarily to oversee the operation and answer questions as they arise, his days begin early and end late. "Breakfast is around 7:30. The clinic is operational by 8:00 A.M. Once the clinic begins to get busy, it will go all day long, usually until around 8 at night. Then there is lab work to finish up and records to update. We had to convince the Malian team to take a day off every week; it's very intense, and the young Malian researchers have incredible stamina. They are there for about six months at a time. We are there for three to four weeks, and at the end, we're all exhausted. They just keep going."

Djimde has a special affinity for this dusty little town. "My dad was in Bandiagara from 1969 to 1972, so I went to school there during my first and second grade." He hopes that his work with drug resistance in *P. falciparum* will benefit the people of his country. Although chloroquine resistance is a prominent feature of the malaria parasites in eastern and central Africa, the malaria parasites in West Africa have remained relatively susceptible to the drug. Djimde, like others, fear this will not last much longer. He believes that much of the problem in Mali comes from the fact that too many people are without health care.

"Chloroquine is widely known as an antimalarial drug, so when people feel symptoms, they can go and buy the drug directly. In many cases, the dosing and interval is not correct, which is a problem with self-treatment. This contributes to emergence of drug resistance."

When asked about his future plans, Djimde is quick to respond. "I can't wait to finish [my doctoral degree], so that I can return to Mali and join the malaria research and training team. I will be carrying on my work with drug resistance. We are reaching the point where we can compete successfully for extramural funding. I think that people are finally beginning to appreciate that it's important to have African scientists involved if they want to succeed. They have to work with the people living there."

Plowe reflects on the present situation with some degree of pessimism: "The only way that we're really going to control malaria is by across the board economic development. Perhaps Roll Back Malaria can assist that, but it is a bottomless pit. [We] have no tools that will get [us] there. Drug resistance is becoming a serious problem, particularly in eastern and central Africa. WHO is sponsoring a series of clinical trials, but there is no way to see that even if they are successful in identifying a clinical regimen, it will be economically feasible. Until you can bring the cost of a treatment down to $0.20 or so, there are not sufficient resources available to have the kind of impact WHO is hoping to have. There are simply no affordable tools."

Donald Ainslie Henderson: Politics and Public Health

Don Henderson is a long-standing proponent of public health and an expert in viral vaccines. He has spent much of his life in the hotbed of international health and politics and came to lead the smallpox eradication program more by serendipity than by design. He remembers the years leading up to his involvement in smallpox from his office in Baltimore with a bit of irony. "It all started in '61 with the measles vaccine. . . . It caused quite high fever in kids . . . so it was licensed with the proviso that it be given with gamma globulin." Measles is a considerable problem in developing countries, with a 5–10% mortality rate in children, and the combination of an expensive vaccine with the need to give gamma globulin made prevention out of reach for most. "The minister of health from the Upper Volta (now Burkina Faso) came to the U.S. and . . . proposed to try [the vaccine] out in kids in the Upper Volta to determine what the effects would be to give just the measles vaccine alone."

The trial was set up in collaboration with the National Institutes of Health, and its outcome was so successful that the six French-speaking countries of West Africa approached USAID for assistance in developing an immunization program. The agency promptly agreed. Their basic idea was good—each country would send a team to the U.S. for training. The teams would return to train additional teams, and USAID would send in the necessary supplies in order to mount the vaccination campaign. When the CDC was asked to evaluate the program with an eye toward taking it over, Henderson dispatched a young Epidemic Intelligence Service (EIS) officer to Africa. "He was there for about 5–6 months and reported that it was a massive disaster—about half the teams had disappeared, AID sent over big vehicles which got stuck in the sand, there was no water to wash hands, no refrigeration. It was a mess."

Henderson recognized a golden opportunity to get his people out into the field. "EIS officers saw health in a different way once they entered the international community and many went into public health as a result." Henderson also recognized that the proposed program was long on political interests, short on public health interests, and unlikely to be sustainable. "This [program] was not entirely philanthropic. The U.S. was looking to establish a presence in the former French colonies." Measles vaccine cost $US 1.75 a dose, and without the ongoing support of AID, the programs would collapse. "The countries in question couldn't even afford yellow fever vaccine at $0.10 a dose."

So he and his colleagues at the CDC hatched a different plan, one they hoped could lead to a sustainable program. Smallpox vaccine only cost about a penny a dose, so they proposed to expand the program to include a block of 18 contiguous African countries and to offer smallpox vaccine along with the measles vaccine where AID chose to provide it. While this made sense from the public health perspective, the expanded program didn't match the political agenda, came with a $US 34 million price tag instead of the $US 6 million AID had budgeted, and left CDC and AID at an impasse.

In the meantime, President Lyndon Johnson was searching for a program he could announce in conjunction with International Cooperation Year at the UN. The CDC arranged for Henderson's plan to be presented directly to the White House, bypassing AID and Henderson's immediate superior at the CDC. Much to everyone's surprise, the White House loved the idea, adopted it, and ordered AID to fund the program.

This was the first contract that AID had awarded to the CDC, but not everyone was pleased by this development. Henderson's surveillance branch chief rightly saw how the program would distort his mission. "My boss was absolutely furious with me. He was in a towering rage. He said, 'This is a huge number of people, it's a big program and will distort everything we're trying to do in our mission. Why did you do this?' " Henderson replied, "It wasn't supposed to be approved!"

In 1966, when the director-general was searching for someone from the U.S. to direct the smallpox program after the World Health Assembly had approved it by the narrowest of margins, there was no better candidate. By then Henderson had managed to set up the measles–smallpox vaccine program in Africa after a mere 15 months. He was ordered to Geneva to begin planning for one of the greatest achievements in public health—the eradication of smallpox.

Henderson's experience has clearly influenced his thinking: "Eradication is not a program to be undertaken lightly. To do so before it is clear that the needed technology is in hand and before the practicability of eradication has been demonstrated in the field is to invite a costly failure and, more important, a loss of professional public health and medical credibility.

"When you get down to it, though, when you're working in a developing country, what can you do with limited resources to really make a difference? Start looking at water, it's expensive to try to do anything, it's slow, it's difficult, it doesn't really work that well in a lot of areas. Try to do sanitation, and putting in sewage [management], etc. It's complicated, it's time consuming, and it's difficult to get the sustained interest of governments. But the vaccines—they're a different story."

Don Henderson received the Presidential Medal of Freedom—the nation's highest civilian honor for distinguished service—on June 20, 2002.

William Foege: Legacy of the Smallpox Campaign

William Foege is in many ways a contrast to Donald Henderson. Where Henderson is direct and brusque, Foege is politic and smooth. Where Henderson paces, Foege sits calmly. Like Henderson, however, Foege has been a powerful force in both national and international public health. He is one of the important legacies of the smallpox eradication program.

Foege's political savvy and passion for his work have taken him from director of the Centers for Disease Control and Prevention, to working with former President Jimmy Carter at the Carter Center in Atlanta, Georgia, to being the lead advisor to The Bill and Melinda Gates Foundation. He has devoted his life to public service, a choice that he attributes in part to his time in Nigeria working on the smallpox eradication campaign shortly after he had completed his medical training. "I actually went to Nigeria to run a medical center. [After I arrived] I got a letter asking if I would be willing to do some work on the smallpox eradication program for the CDC."

The medical center was in the eastern part of Nigeria, an area that experienced a devastating civil war during the eradication campaign there. Foege had just begun organizing the smallpox operation in the area in October 1966 when several small outbreaks of smallpox occurred. He was waiting for the arrival of supplies and personnel, and because his resources were so limited, Foege couldn't take the prescribed approach of vaccinating everyone. So he used a different strategy. When cases were discovered, the team went out and built a "firewall" of immunity around each person, immunizing their family and close contacts, and contacts of their contacts.

The civil war broke out in 1967, within six months of the smallpox outbreaks. The region where Foege was working became known as Biafra. Foege recalls the impact of the civil war on their activities in a recent interview. "That was a horrendous war. You recall the pictures of the children who were starving, the attempts to get food in by airlifts. It was a very difficult time. But they never had a case of smallpox in the area of Biafra during the war, because we were working on the last outbreak the week the war broke out . . . the disease was eliminated just before the war broke out."

With only 750,000 of the 12 million inhabitants of the area vaccinated, transmission had nevertheless been stopped. This was well below the vaccination target of 80% thought necessary to stop smallpox. Foege proposed extending the strategy, which became known as surveillance and containment, to eight other countries. It proved so successful that the new approach was rapidly adopted, and by July 1968, it was the principal strategy for the remaining campaign.

Foege believes that science is often ahead of society's values and that disease eradication offers one way to close that gap to achieve social justice. He offers a quote from Richard Hamming, who said, "If you want to do important work, you have to work on important problems." This has been more than the case for Bill Foege.

Sharon Bloom: When Politics Are Local

Sharon Bloom is a medical epidemiologist within the CDC's National Immunization Program. She believes that her early experiences prepared her well for her work in international health. After graduating from college, she traveled alone through Thailand for a year, visiting both Vietnam and Cambodia. "I like the challenge of being on the ground with no contacts and having to figure out how things work. It is harder—it's based on interpersonal networks rather than the bureaucracy of the health ministry." Her self-confidence proved a critical factor in her role as a STOP (Stop Transmission of Polio) team member in Pakistan.

By 1999, Pakistan's surveillance systems were functioning well, and they had intensified their immunization efforts, but the country continued to report the highest number of cases of polio within the WHO's Eastern Mediterranean Region. Immunization coverage was suspect, but the only way to verify that children were receiving vaccine was to follow behind the vaccinators, checking at every house after they made their rounds. Bloom arrived in Lahore, Pakistan, in January 2000. Her job was to help monitor the vaccinators this district depended on to deliver polio vaccine door to door. What she found was symptomatic of Pakistan's wearying campaign.

For Bloom, getting to know the local politics was essential. "I recall walking in the first day to the district health officer—I was assigned to him because he was not cooperating with the polio mission. The WHO people in Lahore provided me with some intelligence [about him] before they sent me out. They said, 'Oh, he's a tough character. We're sending you there because we can tell you're tough,' and I thought, 'Oh, great!'

"He [the district health officer] was difficult to engage about polio. They have other competing priorities . . . he didn't want to give me attention. I remember having to stand up in front of him and say, 'I'm here to work with you to bring up the [immunization] numbers and the performance of the city to international standards.' I presented a very determined interest and then mentioned that they were not performing as well as Bangladesh." That got his attention!

After she had won over the district health officer, she set to work. "We organized a monitoring program that would make the vaccinators accountable. The surveillance health officer had noticed that some of the local house-to-house marketers used the same system of chalking the houses that the vaccinators used. We hired some of the women who were in the private sector, trained them to do the assessment and give vaccine, and then sent them out into the community to monitor performance."

Chalking houses is a standard practice when house-to-house visits are made. Once a district is mapped and divided into neighborhoods, each neighborhood is assigned to a vaccinator. When a vaccinator visits a house, he or she marks the date and a ratio on the side of the house. The ratio's denominator indicates the number of children under five in the house, and the numerator indicates the number vaccinated that day. If children aren't available, then a supervisor is supposed to go back and try to get them. Bloom remarks, "It's very easy for the vaccinators to mark that they've either gotten all the children or there aren't any children so they can finish early and go home. Sometimes they'll just see the number from the last campaign and write it in without even going to the door. Not everyone sees the need for the campaign. Especially if they haven't been trained to understand why it's so important that each child receive at least three doses of the vaccine—otherwise, they aren't protected.

"The vaccinators claimed 100% coverage, but there was no way to challenge them. So by using these women as monitors, we were able to provide proof that they weren't performing up to standard. They went out after the first campaign and they found about 3,000 kids that hadn't been vaccinated—that's a lot of kids to miss. This, of course, caused a good bit of animosity—[the vaccinators] asked the monitors how they could call themselves proud Pakistanis—but we were able to take this to the district health officer."

Convinced by the numbers, the district health officer brought the vaccinators in and trained them more thoroughly. When Bloom and her recruits went back out to monitor their next campaign, the vaccinators did much better. The number of children missed was reduced by half.

Bloom did a number of other things while she was there to help motivate the vaccinators, including engineering a cash prize for the best vaccinator. She contributed money from her own pocket and persuaded the district health officer to do the same. "The presence of the STOP team adds a little excitement to the campaign. You can see how hard it is to get every kid—it's just so hard. The fatigue is very real. The work it takes to mobilize people and get the media and get people excited. Even the parents will say, 'Oh, my child has had three vaccinations, go away.' You have to persuade them. Plus, people don't see polio as a major threat any more. There are so many other things that are worse, so it's hard to keep everyone motivated at the end."

Phillip Spradling: from Primary Care to the Mountains of Nepal

STOP is not the first international assignment for most of the participating health care professionals, and their experiences are not always positive. Still, they return for more. Such was the case for Phil Spradling, a medical officer within the Vaccine Preventable Disease Eradication Division. Spradling began his career as a primary care physician with Kaiser Permanente but found his practice increasingly frustrating. When his wife was offered a position at the CDC working in AIDS epidemiology, he welcomed the opportunity for a change. Spradling was accepted into the Epidemic Intelligence Service (EIS), the elite team of physicians and other health care specialists who are sent into the field to investigate disease outbreaks around the world.

Spradling's first international experience as an EIS officer took him to Russia. While there, he was mugged, robbed, and brutally beaten. His injuries were sufficient to require that he be airlifted into Germany. Despite his early experiences, he was still anxious to sign onto the STOP team and thus found himself heading for Nepal in January 2001. For Spradling, this was an exciting opportunity, with the added promise of climbing the legendary mountains of this small country. When he reached his destination, however, he quickly found himself caught up in the pressing needs of the eradication effort.

By early 2001, Nepal was well along in its eradication campaign. Spradling spent his first month in Kathmandu, in what he apologetically describes as relative luxury—a hotel with running water and a restaurant. He was assigned to work with the regional surveillance officer to monitor the national immunization days and document vaccination records during the mopping up campaigns. He was surprised at how well their program was working. "There hadn't been a case [of polio] in Kathmandu for quite awhile, but when I arrived, they had experienced a small outbreak of four cases, mostly along the border. There was also a recent case in China that had been traced back to India or Bangladesh. The only way the virus could have gotten there was through Kathmandu. This sparked grave concern. There are remote, sparsely populated areas, especially along the western border, where the population is under the control of the Maoists. You can't be sure that there isn't virus circulating.

"In places like Nepal, there are so many problems, you don't know where to start. They don't have potable water. I'd go into the nicest hospitals and it was just appalling. You'd see the filth and the conditions and the fact that families are herded together taking care of a sick family member—the lack of medications, IV fluid, sterile equipment. I thought it was bad in Russia, but this was just appalling. The local public health officers still seemed truly committed. I was impressed, especially during the mopping up, at how they were able to recruit all these volunteers and supervisors, all of whom were very poorly compensated. Most of what we did was basically cheerleading."

One of the biggest difficulties for Spradling was getting a good translator to help him bridge the language barrier. He went through three before finding one who satisfied his needs. Spradling remembers too many meetings between organizers and vaccinators that would go on all day with his barely understanding what was being discussed. "The translator is your lifeline to whatever is happening. Their meetings were chaotic—not the placid, structured affairs we see here [in the U.S.]. There would be these long and heated exchanges between people in the audience and health officials shouting back and forth. It wasn't in anger—it was just very emotional. I'd say, 'What is this all about?' and [my translator] would say something like, 'Oh, he's just worried that they're not going to get paid.' That would be the summary of this huge long exchange."

Spradling mirrors the enthusiasm of many of his colleagues. "I would definitely go back out. I would certainly do it again. I like the international work, although I would really prefer to be on the frontlines the next time." When asked about the difficulties, he says the worst was being away from his wife for the three months. Upon his return, Spradling got his wife a puppy, perhaps to keep her company when he's away the next time.

He hasn't yet made it into the mountains.

Kathy Kohler: Fulfilling a Dream

Kathy Kohler specializes in epidemiology, the detective work critical to tracking and stopping disease outbreaks. She, like many other STOP members, came from the CDC's Epidemic Intelligence Service. "My EIS assignment was with the polio eradication division, so I was kind of like an insider; I was working with this group already." Before joining the CDC, Kohler chose Emory University for her graduate training "because it was right next to the CDC. I had that in my mind from the time I saw an article about the EIS in a *National Geographic* when I was in college."

Kohler was initially sent to the south of India, to Karnataka, in January 2000. Although polio continued to circulate in high numbers in India's populous and poorer northern region, most of the south was clear of transmission. The exception was this one small region. "The health officials in the national program were anxious to figure out what was going on, so that's where they sent me."

India has a well-developed public health system that extends back to the malaria campaigns decades before. "Those of us who were going to India were walking into a system that works really well, with just a few issues remaining. I was impressed with the dedication of the doctors, who just work tirelessly. I was amazed at the number of hours they put in and their energy and enthusiasm in what can be a somewhat thankless and very difficult job. When you see something functioning that well in a country that is as large and diverse and complicated as India, it's amazing. It's really a model program."

Kohler was pleasantly surprised when she was assigned to a young and dedicated female Medical Surveillance Officer (MSO) working in the Karnataka area. "She was living 10 hours away from her husband and two young daughters. She was taking the train back every Friday night and coming back every Monday morning." Her MSO sometimes found it difficult to travel alone across the broad stretches of India that she had to monitor, and she welcomed Kohler, who would travel and work with her over the next three months. Their job was to make certain that everyone was aware of and trained in how to use the acute flaccid paralysis (AFP) surveillance program, in addition to investigating cases of AFP. "We found a lot of places that didn't really know about it, or didn't have a focal person at the hospital, so we did a lot of consciousness building."

There were some cultural bridges that were hard to cross. Kohler and her MSO were younger women, for example, often talking to older physicians. "The regional health services hadn't really brought in enough of the private physicians and nontraditional practitioners [healers]. The healers in particular, who are called 'quacks,' are outside the loop of the physicians trained in Western medicine. We worked hard to bridge the divide, so that the quacks would feel comfortable contacting a physician and the physicians would not discount their information.

"My days were long, hot, and dirty. [My MSO] was responsible for five districts in the northernmost region of the state. The roads were in poor condition, so it took forever to go from one district to the next. There were mornings when we would get in the car at 6:00 and drive for seven hours to the next district so we could talk with the officials there. We tried to plan our route so we could investigate cases along the way. The distances were just enormous, so I spent a lot of time in cars. One of my most striking memories," Kohler chuckles.

Kohler found other cultural barriers that were contributing to continued transmission of the poliovirus. One such barrier existed in the Muslim communities. "There were rumors surrounding vaccine administration that it might be birth control targeted toward their minority population." Sometimes help can come from surprising quarters, and this was one such time. "While I was there," Kohler recalls, "there was this group of friendly grandmothers who had committed to mobilizing the community. It was so great to see this group of older women who were so proud to be a part of the effort and so committed to making certain that the Muslim children were protected against polio."

When asked about her challenges, Kohler says, "The greatest challenge for me was being so far removed from everything familiar for three months and being away from home. Working kept me busy, it was really interesting and you felt like you were really accomplishing something. I don't know if it was necessarily harder if you're a female. You weren't really inclined to do anything at the end of your day—go out or anything—so I caught up on a lot of reading while I was there."

Kohler has since been back to India three times. First to New Delhi, to work in the national office and then twice into the north, where the polio eradication project is headquartered. Along with many of her counterparts, Kohler sees burnout as one of the most serious threats to the program. "You are talking about 25 million children born every year. The susceptibles accumulate instantaneously. When you think about the number of doses of vaccine to do a National Immunization Day (NID) targeting 120 million children and to do that twice within a month and then do it again . . . it's almost overwhelming. Sometimes parents think this *is* routine immunization."

For India, these efforts will have to continue into the foreseeable future. At the end of 2002, India remains one of three countries where major reservoirs of the poliovirus still exist.

Alice Pope: a Passion for People

An early experience working in North Carolina's immunization program convinced Alice Pope that it was important to wipe out polio forever. "I worked in surveillance and saw a case of polio that was due to the vaccine. Seeing this case of polio really brought it home to me. It was so sad; polio is a preventable disease and then to see that it had been caused by the vaccine. It made this project a very personal experience."

So Pope set her sights on a larger playing field. She entered a training program at the CDC preparing health care workers for international work. The program required each attendee to complete an international assignment, and the STOP program became hers. In January 1999, she became a member of the very first STOP team assigned to Bangladesh, a country of over 130 million people. Bangladesh is a little smaller than Iowa and nestles between India and Burma on the Bay of Bengal.

Bangladesh was more than the site of Pope's first international work; it was her first international travel. "When I was accepted into the program and found out that I would be assigned to Bangladesh, I was very excited and felt very lucky. And I wanted to be very successful."

Pope's passion for what she does is exceeded only by her warmth for the people in Bangladesh. "I had an absolutely wonderful experience on a work level, on a personal level, on a cultural level because of the people I met and worked with. The culture is so warm and open there." In fact, one of the hardest things she experienced was a result of people's openness. "The people are so curious about you. I used to wonder if I was viewed as a princess or a freak sideshow. Hundreds of people would gather around my van and obviously I'm a contrast to their culture." (Pope is tall with cropped blond hair.) "People would get very close and just stand there and stare and stare. I could be there for an hour, and people would still be there."

Bangladesh had an impressive capacity to deliver health services, with a surveillance system in place, and medical officers in all the country's 64 districts and the 464 levels under them. The results coming in for polio, however, were less than impressive. Bangladesh had yet to actualize surveillance for polio, and the virus was still actively circulating at the beginning of 1999. The STOP team, which consisted of four members, joined the country's five polio surveillance officers, almost doubling the capacity.

Pope remembers the political climate in the health care community when she arrived. "The majority of people were very welcoming, although there were a few who felt, particularly at the district level, that this was not the most important problem for Bangladesh. They have all the diseases prevalent in a developing country—diarrhea, tuberculosis, and other respiratory diseases." After contemplating for a moment, she expands her thoughts. "Most people understand. They remember the cases of polio; some people have a real passion for it because they feel like there is so much they can't control, but here is something they can actually do."

Pope herself had some early concerns about being a part of a single-disease program. Her apprehensions were immediately laid aside. "It's very rare to hear anyone [in Bangladesh] get into a big discussion about what they aren't doing. In fact, most people agree that they wouldn't have the resources to do anything about the other issues even if they weren't involved with the polio project. Plus, there's so much heart here [at the CDC] from the top with Steve Cochi [the director of CDC's Global Immunization Program] all the way down. No one wants this to be a program that just comes in with a single purpose and leaves nothing behind."

If anything, Bangladesh's program has become a model for capacity building. The polio eradication program incorporates other preventive measures such as the distribution of vitamin A during an NID. Their surveillance system tracks other diseases such as measles and neonatal tetanus, both of which are targeted as part of a broader immunization program. The national health officials and WHO representatives operate very closely as well. As Pope describes it, "The office for polio and the EPI [WHO's Expanded Program for Immunization] is at the same government building. It's so well integrated that most people don't distinguish between who is WHO and who is government." She continues, "People who are working have gotten great skills. They're the ones who will be there to support those other programs when we're done. They'll be the ones who will make the difference. In the end, there will be a huge benefit left behind."

The greatest challenge in Pope's view for Bangladesh is maintenance. The country hasn't had a case of polio caused by wild-type virus since August 2000. "We're all really hoping that that is accurate. We're doing a lot of work to make sure, but sustainability is a big issue. People talk about being burnt out, but they're doing it."

Pope joined the STOP program at the CDC permanently at the end of her first trip to Bangladesh. She has since returned to the country three times to supervise and coordinate other teams. A part of what continues to motivate her is her admiration for her colleagues. "I've never worked with such a group of committed, rational people. There's a real belief that what we are doing is really going to help other people. I don't know how they hold onto that passion, but they really seem to. Steve Cochi has really incorporated that belief into his life, into what he does and how he touches other people. It's delightful to be a part of it."

Duane Kilgus: from Desk to Desert

As the Project Officer for Cooperative Agreements and Contracts, Duane Kilgus spends most of his time at the CDC behind a desk. But that's not how he started; Kilgus joined the CDC after spending a part of his life working with Native Americans on reservations in New Mexico. It was an experience that he equates to being in a foreign country. "That's what I wrote down when I applied for a spot on the STOP team. All the same things apply. You have to deal with a different culture, a different language, a different way of doing things."

His experience was enough. In September 1999, Kilgus left for Quetta, the provincial capital of Baluchistan. In some ways, this area of Pakistan is like the deserts of the western United States where Kilgus worked before coming to the CDC. Baluchistan is large, and its landscape is barren and hostile. In contrast to its vastness, it is home to only a little over 4% of the country's population. It is by far the least developed of all of Pakistan's provinces by almost any measure, be it basic infrastructure and housing, per capita income, literacy rates, or health care. Perhaps even more significantly, Baluchistan shares its longest border with Afghanistan to the north and Iran to the west.

When Kilgus left for Pakistan, his stated objective was to improve surveillance. Once there, however, he found himself cast in a slightly different role. "The fact that you were there as a representative of the WHO meant that this was important to the rest of the world. That gave an extra spark that reinforced the idea that it [polio eradication] needed to be done."

Kilgus was soon working on the front lines. "I put on my salwar kameez [the traditional Pakistani dress] and went out into the public and gave vaccine. I stopped at all the different sites and helped with training health workers." Kilgus also helped with social mobilization—getting information to people so they knew that a National Immunization Day was coming. "The whole idea of television, banner signs, electronic media, they don't happen there. Renting a camel cart with a loud speaker is about the only thing available except mobilizing the message through the mosques. And amazingly, it works—we got 84% coverage when I was there."

At a typical stop, when the announcement was made that the immunization team was there, Kilgus found himself immediately surrounded with parents and their children. "There was no concept of lining up in an orderly fashion. I was kneeling so I could hold the children safely, and I finally had to stand up so that I could just breathe. They wanted their children to be healthy. I think the desire to have healthy children—that crosses all kinds of borders, races, and ethnicities. Have you ever heard of any other situation in the world where they could negotiate a ceasefire just so they could vaccinate children?"

Working on a STOP team in a Muslim country does present some challenges, especially during house-to-house vaccination efforts. "When the woman is home, she talks to you through the door. You never see her. But everyone knows about polio, so they are happy to answer questions about their children's status. I had one really cool experience where I was training a group of women with a Pakistani physician. The physician had to leave for some reason, so here I was staring at a group of 18 or 20 lady health workers. They spoke a little English so we could communicate. I started asking questions about the kinds of problems they had experienced. This was very different for them. Most of the time the training is very formal. If questions are asked, it is in a very formal manner. So when I started asking for their opinion, they all wanted to get their opinions out. As soon as they felt safe, they started asking lots of questions and offering their views. Then when the doctor walked back in, everything just shut down."

Kilgus, like many of his colleagues, was anxious to return after his first trip. So he was ready to leave for Pakistan again after completing his training on September 21, 2001. The work would be even more important this time; he was to help during an NID that would be conducted with Afghanistan and Iran. By coordinating the three countries' immunization efforts on the same day, the team hoped to get to the children of the refugees who move constantly across the northern border. STOP 9 was canceled in the wake of the attack on the World Trade Center and the Pentagon and the WHO and international health care workers withdrawn. Kilgus reflects upon the impact subsequent events have had on the polio program in this region of the world. "I think that the people who are there are dedicated to it, and I think they will stick with it. I don't know if it will be as good because they won't have the people from WHO to help, but they are very capable and they've been doing it for a long time. I think they'll continue with the scheduled day, and you know, be on time."

When asked about how he felt when the training was cancelled, he replied "I went back to Pakistan in February . . . to help with refugee work on the Afghanistan border, and I'll go back again as soon as I can. This last time was exciting for a different reason, because I started seeing the numbers and seeing that we could realistically wipe polio out by the end of this year or next year, and that was really exciting. Being involved with the eradication of a disease is something that most people never get to be involved with. It's just really neat to be involved, and the idea to go back and finish it, is just really neat."

Virginia Swezy: a Champion in the Final Leg of the Race

Virginia Swezy perches on her chair in her office, surrounded by pictures of children from around the world. Swezy, now the deputy director for STOP, got her first taste of working for international health in Togo, West Africa, as a member of the Peace Corps. She didn't join the STOP program until shortly after the first teams went into the field. She regrets missing the excitement of being there at the beginning, but even three years later she rarely has a free moment. Swezy is the person responsible for maintaining organization and order. If Linda Quick, the program's director, is the heart and soul of STOP, Swezy is its hands and feet. "For the first year and a half, it always felt like the first time. There was always something that jumped up to surprise us. Now, we've gotten into a routine. During the first year, we were working ten- to 12-hour days and every weekend."

Swezy wasn't following a master plan when she got involved with the program. She was doing an internship in the Sexually Transmitted Diseases Division of the CDC, but she's always had a passion for working with children. When the invitation came to join the STOP program, she accepted immediately because it brought her back to a program with children at its center. She knew it was a place where she could make a difference.

"When you look at polio, it's such a tiny little thing on the plate. So you can go to the district hospital or clinic, and polio just isn't the biggest problem. You may be there to talk to them about polio, but they have 300 mothers lined up with children who are dying from malaria. Still, one of the beauties of polio eradication is that it is very easy to tackle. If we can tackle it, we are setting up a system that can be directed to other issues in the future that they care more about. The infrastructure and surveillance system in place with their cadres of trained people are resources of the polio effort that will remain behind when polio is gone."

Swezy's coordinating efforts can be a delicate balancing act. "We work behind the scenes. When we are forming teams, we go to WHO and to the regional offices to let them know we are preparing teams to go out. They help us identify opportunities and we evaluate them using a couple of key characteristics. There must be strong supervision for our team members. It doesn't work just to put someone out into the field with no supervision. We're also very concerned about safety issues. There's a lot of thought and a lot of consideration before we put someone out into the field, and we monitor the political situation until the team leaves. We even have had to pull people out on occasion. For example, one team member had to be evacuated from Afghanistan when the Taliban came in reputedly to search for someone. It's delicate, because if a country requests a team and we have to refuse because of security issues, we don't want to compromise the relationship."

STOP's role in the larger scheme of the effort that began in 1988 has been small, but it has been a giant in the final months of the eradication campaign. Swezy sees it as a matter of numbers. "We can put lots of people-years into the field over a short period of time. Putting 40 people into the field for three months adds ten people-years to the effort."

Although she spends much of her time coordinating STOP's activities from Atlanta, Swezy really loves going out, rolling up her sleeves, and getting to work where things are happening. "When the NID happens, it's everything that you've been working up to, and to see that happening is really special." But that reward doesn't come easily, and burnout remains a central issue for the campaign. "Everything is ten times as hard in the developing world. Just getting from place to place can be overwhelming. The roads are just awful and a day out is physically draining. They don't have refrigerators that work or resources in their clinics. They're often responsible for big regions and have to meet with and train health care workers that are as tired as they are." Then there's the added burden of working in regions where conflict exists. "In places like Angola and Nigeria, surveillance is extremely difficult. If you're trying to cover every district, people are at risk. People are scared and going out every week into districts that are insecure, may or may not be friendly—how do you ask people to do that? Still, people go."

Swezy knows as well as anyone that many challenges still remain, but her admiration for the people who have worked in and led this campaign inspire her every day. "I have to remain optimistic, and I am. Certainly there are days when we get depressed, but we're so close to ensuring that no child anywhere, regardless of race, ethnicity, or socioeconomic status will ever have polio again. We just have to maintain the push. We just can't let go of it."

Ultimately, she knows all the effort is for the world's children, and that's why it matters. "There really is nothing more important for us to give to the world than to wipe out a disease from the planet. It's been a huge struggle and we're so close that if we just maintain our commitment to seeing it through, then it's a gift that we have forever. That's a historical moment in itself, but we're also setting the stage for countries to take on other challenges in the future. That's an important legacy."

Steve Stewart: Roads and Rivers

Steve Stewart's early experience working in a developing country won him a spot on the STOP team heading for Dhaka, the capital city of Bangladesh. Stewart began his career in the 1970s with the Peace Corps in Sierra Leone, West Africa, and after 20 years as a health communication specialist, he wanted to get back out into the field. But that wasn't his only reason for going. "Honestly, one of the other attractions of working in the polio campaign is the fact that there is a very tangible goal—eradicating a disease. Certainly in public health and medicine in general, that is very unusual. Usually you're trying to reduce something or prevent something rather than get rid of it all together."

Stewart's assignment took him away from Dhaka through miles of flat and densely populated countryside. With a smile, he compares the experience of driving along the two-lane roads there to working in a conflict country. "You're sharing the road with donkey carts, bicycles, motorcycles, cars, trucks, buses, all of which are going at different speeds, and some people are passing one another all the time. I can't tell you the number of times I looked up and saw a bus in the middle of the road passing a bicycle and the truck we're approaching passing a cart and we're heading right at each other. It's a very common thing to swerve and miss each other by a foot or two and you're traveling along at 60–70 miles per hour. When it was time to leave, we all just gave a sigh of relief that we had survived the roads in this country."

Stewart credits some of the success of his tour to his translator. "Biplob [which means 'revolution' in Bangla] was someone who was very good at talking with all kinds of people. We got to know each other pretty well. . . . We had both come from relatively poor families and had to work hard as kids. He got scholarships to go to medical school, and his parents worked hard to be able to put him through school. I think in part because of his background and in part because of his personality, he was able to speak easily and be comfortable with people from a variety of backgrounds, from the poorest people we met to the civil servants that were high up in the health bureaucracy."

Stewart remembers one experience with Biplob out of the many during their travels around the country together. "When we were doing monitoring for the NID, one of the places in our district was a very large bordello, which was almost like a village in itself, near the ferry landing. There were probably 3,000 to 4,000 people living there. Because of the density of population, we had to check to see how well the NID had gone. Biplob is Muslim. As we pulled up I could tell he was feeling sort of anxious. This was unusual for him, so I said, 'What's going on?' He said, 'Steve, I fear this place.' I laughed and asked whether he was concerned that people would think we were going there as customers and he said, 'Yes, exactly.'" Stewart immediately had a solution. "We had our yellow vests and caps identifying us as working in the polio campaign, so I said, 'Why don't we put them on?' and he snapped his fingers and said 'Ah, good idea!' Within a couple of minutes of being there, he was chattering away with the mothers about the immunization day."

It's sometimes difficult to describe what it means to immunize the world's children against this disease. The immensity of the polio campaign was brought home to Stewart during his first ride on the ferry across the Padma River back into Dhaka. "I was up high on the ferry. For the first time, I got a longer view of what the landscape was all about. Just being up high, I had a sense of [how] vast this country is. Just to see the very small pockets of villages and people along the river in places that clearly would be difficult to reach, I got a sense of the real challenge, the logistical challenge of reaching all the kids with the vaccine."

The ferry itself would become a safe haven for Stewart after a long week spent doing surveillance in the countryside. "There was something about being on the river, coming back toward Dhaka. Having worked in the field all week, sometimes under pretty challenging conditions, and having to be polite and diplomatic, to get on the ferry was like entering another world for a little while. It was very peaceful."

Stewart reflects that his greatest personal challenge was working through the cultural differences between the "get down to business" approach of the West and the need for social formalities before business in the East. "It really is about establishing relationships with people. You go in and have several cups of tea and meet everyone who works at the facility. You have to slow down a little bit, defer to the people who run the show locally, and explain carefully and clearly what your role is. You're not coming in to take charge or show anyone up, but to assist with the work locally and to get their input on what the challenges are—what needs to be done differently, what's working well, where they need extra help, whether they have the materials they need, whether the people are trained, whether there are any newly hired people who might need training.

"There are times when you just sort of grit your teeth and say this is hard. I feel frustrated today, I'm staying in a place where the electricity keeps going off, and I wonder if I'm going to get a decent meal tonight. I wish the fan actually worked in this room. Or you're just hot and tired and away from home and it's hard to communicate with folks back home. But you forget about that once it's over. When I look back on it, I would say I felt stretched, and that I grew professionally. For someone who spends a lot of his time working in front of a computer, it was a real pleasure to get out in the field and participate in something like this. That's why if I were asked to go again, I would definitely go back. I would go back there, I would go to another country, gladly."

Fabio Lievano: Opening Pandora's Box

Fabio Lievano, a native of Colombia, is a medical epidemiologist in the Global Measles Branch of the CDC, but he first cut his teeth on international work in Bosnia. "I was working five years in Bosnia with the WHO during the war. I was running immunizations in Bosnia, so I have a lot of background doing these kinds of programs during wartime."

His experience working in a conflict country, however, didn't prepare him for what he found when he arrived in Chad in May of 1999 as one of the three-member STOP team. Lievano describes his experience in no uncertain terms: "Chad has been the most difficult experience I've had in my life, much more than Bosnia. In Bosnia, we were in the middle of a war, in the middle of winter, and it was very cold, no electricity, no water, very little food, but Chad is very, very hot . . . the food, it's not easy to find a restaurant—actually there are no restaurants in many places, it's dusty, there's no water. It's difficult. There are snakes. There are a lot of things. It was only three months of my life, but believe me, the first week I arrived there I wanted to turn around and come home."

Although many of the remaining endemic countries welcome assistance from the international community, Chad would prove an exception. Lievano was confronted with that reality immediately upon his arrival. "We were expecting a big reception because we believe we are coming to save the world or something like that," Fabio chuckles. "We arrive at the airport and the first surprise is, there is no band here. Actually, there was nobody. We were waiting there and no one arrived." Lievano remembers turning to his colleagues and saying, "Oh my goodness, maybe this is not the right country. We thought it was Chad, but maybe this is another place!"

The reality was that in Chad in 1999, polio eradication was hardly a top priority; in fact, it wasn't a priority at all. "I was thinking that the first priority for the country should be polio eradication, and then when you arrived there, you see that you are not even the last priority. They ignored you, they don't want to discuss much about polio." The local WHO staff were under enormous pressure to concentrate their efforts on systems to provide general health care, and surveillance for polio was almost nonexistent. This left the impression that polio itself was nonexistent. Lievano reflects on the impact, "They were saying, 'We don't have cases of polio. Why are you sending three people to demonstrate that we don't have cases of polio? Believe us, we don't have any, we eradicated polio a long time ago.' "

Lievano didn't take well to being ignored, and he soon found a way to get the WHO staff's attention; he threatened to leave. "At the beginning, because we threatened to make the situation worse for them—they would have to explain why we left the country—they provided us with a vehicle, translators, a driver, and then they sent us to the worst imaginable places on the earth!" The team divided into three groups and began to canvass Chad for cases of polio. "Among the three of us, we covered 40,000 kilometers. And the roads are very African roads. I always thought that 40 kilometers per hour was the slowest a car could drive. Actually you can be even slower. Sometimes there were no roads, and they have to invent the roads. When it starts raining, the cars get stuck, or if there is a lot of sand, the car can go into the sand and it will be difficult to get it out. To drive to anyplace in Chad, it was a question of one or two days—anyplace."

"I was sent to the south first, close to the border with Cameroon. The last city in Chad before you enter Cameroon is Sahr. The first day I arrived I was really depressed. It was a small place, raining, no electricity, the water is there, but there is a lot of cholera and you aren't sure that you want to drink that water. By the time I arrived in Sahr, I had diarrhea, and I had diarrhea the whole three months I was there. It's hard. You feel alone, you feel like maybe you are not doing the right thing because the people are not that interested in the program. But then after the third day, you start to look for cases, you start to do the real work that you were sent for, and then is when you open the Pandora's box."

It didn't take long for the teams to establish that Chad indeed had a problem. Lievano exclaims, "They have polio everywhere—they don't know about all these cases, so they don't feel they have a problem, but to make a comparison, they reported only two cases of polio in 1998. In 1999, after we leave, they reported 110 cases of polio in only three months. You see the magnitude of the problem. Chad is a huge country in terms of geographic area, but in density of population it is a small country, and one of the highest if not the highest incidence of wild poliovirus in the world."

And it wasn't just the number of cases that signaled trouble. It was the genetic diversity among the polio strains. The latter told the team that poliovirus had been circulating there unabated for many years. Lievano explains, "Usually, the same family of poliovirus concentrates in one country. In Chad, there were all sorts of wild virus families in circulation. It says that they have been an endemic country for many years. Whatever they were doing, they were not doing it right because they have too many cases of polio."

Lievano traveled next to the north, an area of the country still populated by guerrilla groups and civil unrest. He remembers one particularly harrowing ride. "One day we were traveling north toward Sudan, and we found mines on the road. I was so scared, and the driver just told me, 'Don't worry. I just will avoid them.' These were antitank mines. We were driving in a Land Cruiser. These antitank mines will really destroy a tank. Imagine what they will do to a 4 by 4 small jeep!"

Lievano has personally seen how a program like polio eradication can have an impact on a country's capacity to deliver health care. "[The WHO staff] say they want to do something to develop primary health care or immunizations at the same time, tuberculosis, HIV, we want to improve water. But that is totally unrealistic. How much money would we need to do something like that? In Bosnia we have a lot of support from U.S., U.K., Sweden, and other countries, and we developed a primary health care project that was very comprehensive. We were asking for some millions of dollars, not that much, and we got no money and it was a big failure. To develop primary health care in a developing country is a complicated issue. Rather, I would choose a program like polio or measles to build upon and help other programs in primary health care. We get more money that way, we see results quickly, and we can prove to the doubters that this is working and besides, we are helping other programs."

Lievano has been back to Africa many times since his first visit—to Chad, Angola, and the Democratic Republic of the Congo. Yet he doesn't see himself in the same light as the people he has found there. "The great people you find in the middle of the bushes. There are many missionaries who have done a lot for the community. When you come with the polio program, they immediately say yes, we support you 100%. . . . No one has come, ever, to this place. . . . They are not recognized, but there are hundreds of them, and they are the heroes of the eradication program. They sometimes pay from their pockets just to send samples to the capital. They receive no salary; they have five dollars and they spend one. If those are not heroes, I don't know what is."

Donald Hopkins: the Blowing of a Certain Trumpet

Ralph Henderson, the assistant director-general of the World Health Organization, has a favorite quotation that he uses when he speaks about leadership. It comes from 1 Corinthians: "For if the trumpet give an uncertain sound, who shall prepare himself to battle." His colleague, Don Hopkins, knows the importance of a certain trumpet when it comes to an eradication campaign. Beginning with his earlier days as deputy director of the CDC, he has led a variety of initiatives on the ground and from behind an administrative desk, and he appreciates all too well how easily programs can go awry when leadership falters.

Hopkins has also seen the positive effects firsthand that disease eradication can have on people's lives in regions with struggling economies, and that is sufficient to motivate him to action. Hopkins is realistic about political realities, however, and he is quick to point out other benefits. "We really are a global village. Anything we can do to raise the economic level and help people develop self-sustaining economies expands the global market."

Guinea worm disease is an affliction suited to such a call for action. Hopkins was in India when he saw his first case; the worm was just beginning the painful course of emerging from a man's knee. Over the next few years Hopkins witnessed similar suffering again and again: worms emerging from the back of a child's head, from under a man's tongue, even several worms pushing their way through the skin of a single person. So when he was invited to join the Carter Center in Atlanta, Georgia, to lead their guinea worm eradication initiative, he couldn't refuse.

A veteran in the campaign to eradicate smallpox in West Africa, Hopkins was well suited to the role, and he was determined that guinea worm would follow smallpox into the disease museum. His task was made harder in the beginning because WHO actively resisted proposals for eradication of guinea worm. There were questions about the priority of guinea worm when other health needs were so great. There were also questions about a campaign that targeted a single disease in countries that needed so much work on the delivery of health services. And it was accepted among the international health community that no disease could be eradicated without the full support of the WHO. Hopkins forged ahead anyway. He knew that without WHO's support, he would have to convince reluctant governments to spend part of their meager health budgets on guinea worm eradication. Hopkins began to make the case that guinea worm disease was sapping their nations' economic strength. Because most health ministers were focused on the more obvious health problems in the major urban centers, they had no idea that guinea worm, a disease occurring in their most remote villages, was a problem. Hopkins pointed out to the leaders of Nigeria that the country was losing millions of dollars in cocoa crop revenues, the second biggest foreign currency earner besides oil, because in many small villages 60% of the farmers could not work during worm season. He argued to the leaders of Ghana that education and agriculture were the backbone of village development. Both were under threat from guinea worm. Men and women cannot work and children cannot go to school in the worm season.

Now, as associate executive director of the Carter Center, founded in 1982 by former President Jimmy Carter, Hopkins is closer than ever to his goal. Both men share the conviction that economic growth in the developing world has to be accompanied by improvements in public health and eradication of specific diseases. Under the leadership of Hopkins and Carter, programs in the endemic countries grew. Beginning with efforts to improve water supplies as a subgoal of WHO's International Drinking Water Supply and Sanitation Decade, the two men soon identified a new strategy and a new partner. The parasite that causes guinea worm disease contaminates the shallow pools that supply drinking water for many of the villages in endemic countries. The new strategy called for running the contaminated water through a special nylon filter before drinking it. With this simple procedure, the villagers could prevent the disease completely. And because humans are a key part of the life cycle of the worm, when transmission is stopped, the parasite will be eradicated. The new partner was E. I. du Pont de Nemours & Company. The company has supplied the nylon and made the filters for over a decade.

Hopkins is not reluctant to challenge those who argue against targeted programs such as guinea worm eradication. "People in these neglected communities need help. I have yet to visit an African village endemic for dracunculiasis [guinea worm disease] that is suffering from too many visits by health care workers from different programs. The real problem is getting any health services to such communities. In the broad benefits it has provided and in its support of the public health staff and volunteers who are producing those benefits, one can assert . . . that in addition to eradicating dracunculiasis, the Dracunculiasis Eradication Programme has done more to improve primary health care in endemic communities than many primary health care programs."

The number of cases of guinea worm began to fall as country after country joined the campaign. As success followed success, the WHO finally made eradication a goal of its own in 1991. Since that time, the number of cases of guinea worm disease has dropped from more than 3 million to less than 100,000. Over 90% of the remaining cases are in only five countries—Burkina Faso, Ghana, Niger, Nigeria, and Sudan. With any luck, guinea worm disease may even precede polio into extinction.

Walter Dowdle: the Leader People Barely Know Exists

Walter Dowdle peers intensely through his wire-rimmed glasses as he quietly discusses the pros and cons of eradication as a public health strategy from his office in a peaceful suburb of Atlanta, Georgia. The office is at the home base of The Task Force for Child Survival and Development, a nonprofit public health organization whose mission is to help public and private organizations achieve their goals in health promotion and human development. The task force, itself a product of a collaboration between WHO, UNICEF, the United Nations Development Fund, World Bank, and the Rockefeller Foundation, operates by building coalitions, forging consensus, and leveraging scarce resources. Dowdle is well matched to this environment and to his role, for he has spent much of his professional life in public health working to achieve consensus. His years at the CDC as deputy director and then acting director and finally at WHO have prepared him as one of the consummate statesmen within the public health community.

Dowdle is committed to the idea that disease eradication plays a singular role in global public health policy. In that regard, he places the value of eradication versus control in no uncertain terms. "My definition of control is whatever the public will let you get by with. Control is not really a good public health goal. The ultimate goal is getting rid of the disease. That's the ultimate goal in public health. When you talk about eradication, you're really just talking about when, it's not a matter of whether that should be the goal, it's just a matter of when. So whether you decide to do it over 20 years or 30 years by raising the standard of living, much as we got rid of many diseases in this country, or whether you do it methodically over a long period of time, or whether you do it with a special program, the goal [of public health] is to get rid of disease."

Among his varied roles, Dowdle serves as a senior WHO consultant to the Polio Eradication Initiative. From his vantage point, there is no reason to doubt that wild poliovirus transmission will be stopped. "I don't think eradication is the issue. I think we have shown very clearly that the virus can be eradicated. And nobody showed that malaria could be eradicated, nor could the vector be eradicated. But we have shown over and over that polio can be eradicated. So that is not an issue. It's not even a question. The only reason for failure is if the supporters just withdrew, but can you imagine that happening? Can you imagine the responsibility that people feel to keep that from happening?"

Although he has been a champion of both the polio and guinea worm programs, Dowdle urges caution in considering another effort any time soon. "When is it appropriate to enter into a special program? That's a different question. There are a lot of things that have to come together. The time just has to be right to do it. First of all, the disease has to meet all the biologic criteria. There must be the international will to do it. There must be somebody behind it in terms of really championing the program, of being a cheerleader for the program."

That person may be Walter Dowdle, but if it is, and when his work is done, his colleagues will almost certainly say, "We did this ourselves."

Steve Cochi: Turning Dreams into Reality

Steve Cochi is the director of CDC's Global Immunization Division. His role as a visionary leader has been central in inspiring the people who work with him, making certain that no child has been left behind in the current initiative to eradicate polio. Even so, he wishes that the campaign could have gone faster. "Looking back, it seems a shame that the global initiative was launched in 1988, but momentum didn't really develop until 1995. After the Americas, the next region that went forward aggressively was the Western Pacific Region, but that didn't really start until 1993 when China committed to doing National Immunization Days. Then it wasn't until 1995 that the Southeast Asian Region started implementing the proven strategies and 1996–97 before Africa started gearing up. We might have reached this target (polio eradication) quite readily had the momentum, commitment, and resource mobilization taken place earlier."

Cochi believes that part of the barrier was simply inertia in getting such a massive program under way. But there were other factors as well: "There were a lot of skeptics, not only people who didn't believe that it could be done, but also people who ideologically didn't believe that this was an appropriate use of resources. There were many obstacles that had to be overcome. WHO itself wasn't entirely convinced that this was a priority and something worthy of being pursued and having their reputation put on the line."

He gives credit to a lot of people in getting the program under way and making certain that it didn't falter. "At the beginning, it was Ciro deQuadros. He had the vision and the courage to mobilize an entire region, test the strategies, and demonstrate that they worked. By 1991, that was in the bag. In spite of that, there was tremendous inertia for the next five years. So Ciro carried that water a long way, but it had to be carried much farther after 1991. No single individual drove this—it had to be a symphony of voices and partner organizations that came together to make this happen. Rotary has been the soul of the program. Without Rotary keeping the pressure on, keeping the advocacy going, the program would have been much more difficult to maintain. They were a factor in persuading the WHO to go global."

Cochi has also been in the middle of the debates about the best way to spend limited resources. "I believe that the polio eradication campaign has changed some opinions about vertical programs. By any measure, this [the eradication campaign] has been successful; it's brought not only immunization, but also health higher up on the agenda of many countries. The polio campaign has been very visible, it's concrete, and you can measure its success. It's made people feel it's possible to come up with an ambitious target and succeed.

"I think this discussion about horizontal versus vertical will go on for eternity. I don't think it [the polio campaign] will change the opinion of many people who are very health systems/development oriented and see any targeted program like this as a threat to building the system. It's a way of thinking that I don't think will be eradicated just because a program like the polio eradication program has been successful. I do hope that the world will move more toward acceptance of the value of having both—targeted programs to give people goals and objectives to strive for as well as attention to building the system under which those programs need to operate. They can each help each other out and I hope that we'll have a better recognition of that in the wake of the polio eradication program so there is less of this either/or kind of discussion, which doesn't serve anybody—especially the children who are supposed to be the recipients of these programs."

Eleanor Roosevelt once said, "The future belongs to those who believe in the beauty of their dreams." If she was right, then Steve Cochi will have an important place in the future of disease eradication.

The World Health Organization

On the tree-lined Avenue Appia in Geneva, Switzerland, an imposing building stands in the middle of an expanse of lawn. The flags of many nations fly from a series of tall flagpoles surrounding the building. The structure houses one of the most influential players in the global battles to improve the health of all the world's citizens. The building is the headquarters for the World Health Organization, also known as the WHO.

WHO traces its beginnings back to the mid-1800s, to the first International Sanitary Conference held in Paris. The conference participants attempted to create an international convention that would bring infectious diseases under control. It failed in its mission, but it brought together a group of like-minded individuals with the idea of global strategies to fight diseases. The League of Nations, founded in 1919 and charged (among other things) with issues related to control and prevention of infectious diseases, came a step closer to setting a global health agenda. Ultimately, it was 1948 before the necessary world powers came together to form the WHO.

As a specialized agency of the United Nations, WHO today is the directing and coordinating authority on international health work, bringing together representatives from its 191 member states. Although its central administration headquarters is in Geneva, much of its effectiveness derives from its six regional offices representing Africa, the Americas, Southeast Asia, Europe, the Eastern Mediterranean, and the Western Pacific. This decentralization helps to support and monitor activities on a regional level, giving the WHO many opportunities to appreciate country-to-country differences that affect programs and, subsequently, to adjust to those differences.

The WHO's policy-making body is the World Health Assembly (WHA), which meets annually in Geneva. The WHA includes delegations from each member state, and its purpose is to authorize the WHO budget and decide on major policy matters. All eradication efforts must be approved by this governing body, which theoretically serves to ensure the commitment of all nations, as much as is possible, to the social and financial realities of a major campaign against a specific disease agent.

Bibliography

Malaria

Bremen, J. G. (ed.). 2001. The intolerable burden of malaria: a new look at the numbers. *American Journal of Tropical Medicine & Hygiene* **64**(Suppl.) January/February 2001.

Carson, R. 1962. *Silent Spring*. Houghton Mifflin, Boston, Mass.

Centers for Disease Control. 1986. *Plasmodium vivax* malaria – San Diego County, California, 1986. *Morbidity and Mortality Weekly Report* October 31, 1986.

Clyde, D. 1967. *Malaria in Tanzania*. Oxford University Press, London, United Kingdom.

Colbourne, M. 1966. *Malaria in Africa*. Oxford University Press, London, United Kingdom.

Desowitz, R. S. 1981. *New Guinea Tapeworms and Jewish Grandmothers: Tales of Parasites and People*. W. W. Norton, New York, N.Y.

Desowitz, R. S. 1991. *The Malaria Capers: More Tales of Parasites and People, Research and Reality*. W. W. Norton, New York, N.Y.

Farid, M. A. 1980. The malaria programme—from euphoria to anarchy. *World Health Forum* **1:**8–22; discussion **1:**22–33.

Farid, M. A. 1998. The malaria campaign—why not eradication? *World Health Forum* **19:**417–427.

Gladwell, M. 2001. The mosquito killer. *The New Yorker* 2 July 2001, pp. 42–51.

Gramiccia, G. 1981. Health education in malaria control—why has it failed? *World Health Forum* **2:**385–393.

Harrison, G. 1978. *Mosquitoes, Malaria, and Man: a History of the Hostilities Since 1880*. E. P. Dutton, New York, N.Y.

Litsios, S. 1997. Malaria control, the Cold War, and the postwar reor-ganization of international assistance. *Medical Anthropology* **17:** 255–278.

Litsios, S. 2000. Criticism of WHO's revised malaria eradication strategy. *Parasitologia* **42:**167–172.

Marshall, E. 2000. A renewed assault on an old and deadly foe. *Science* **290:**428–430.

Molineaux, L., and G. Gramiccia. 1980. *The Garki Project: Research on the Epidemiology and Control of Malaria in the Sudan Savanna of West Africa.* World Health Organization, Geneva, Switzerland.

Oaks, S. C., et al. 1991. *Malaria: Obstacles and Opportunities.* National Academy Press, Washington, D.C.

Packer, R. M. (ed.). 1997. Malaria and development. *Medical Anthropology* **17**(special issue).

Phillips, R. S. 2001. Current status of malaria and potential for control. *Clinical Microbiology Reviews* **14:**208–226.

Rice, R. 1954. DDT. *The New Yorker* 17 July 1954, pp. 31–56.

Spielman, A., and M. D'Antonio. 2000. *Mosquito: a Natural History of Our Most Persistent and Deadly Foe.* Hyperion, New York, N.Y.

Taubes, G. 2000. Searching for a parasite's weak spot. *Science* **290:**434–437.

Tishkoff, S. A., et al. 2001. Haplotype diversity and linkage disequilibrium at human G6PD: recent origin of alleles that confer malarial resistance. *Science* **293:**455–462.

Vogel, G. 2000. Against all odds, victories from the front lines. *Science* **290:**431–433.

Warshaw, L. J. 1949. *Malaria: the Biography of a Killer.* Rinehart & Company, New York, N.Y.

Wills, C. 1996. *Yellow Fever, Black Goddess: the Coevolution of People and Plagues.* Addison-Wesley, Reading, Mass.

Smallpox

Basu, R. N., et al. 1979. *The Eradication of Smallpox from India.* World Health Organization, New Delhi, India.

Cockburn, W. C. 1966. Progress in international smallpox eradication. *American Journal of Public Health* **56:**1628–1633.

Crawford, D. H. 2000. *The Invisible Enemy: a Natural History of Viruses.* Oxford University Press, Oxford, United Kingdom.

Durbach, N. 2000. "They might as well brand us": working-class resistance to compulsory vaccination in Victorian England. *Social History of Medicine* **13:**45–62.

Fenner, F., et al. 1988. *Smallpox and Its Eradication.* World Health Organization, Geneva, Switzerland.

Foege, W. H., et al. 1971. Selective epidemiologic control in smallpox eradication. *American Journal of Epidemiology* **94:**311–315.

Henderson, D. A. 1976. The eradication of smallpox. *Scientific American* **235:**25–33.

Henderson, D. A. 1978. Smallpox—epitaph for a killer? *National Geographic* **154:**796–805.

Henderson, D. A., and F. Fenner. 2001. Recent events and observations pertaining to smallpox virus destruction in 2002. *Clinical Infectious Diseases* **33:**1057–1059.

Henig, R. M. 1994. *A Dancing Matrix*. Vintage Books, New York, N.Y.

Hopkins, D. R. 1983. *Princes and Peasants: Smallpox in History*. University of Chicago Press, Chicago, Ill.

Hopkins, J. 1989. *The Eradication of Smallpox: Organizational Learning and Innovation in International Health*. Westview Press, Boulder, Colo.

Imperato, P. J., and D. Traore. 1968. Traditional beliefs about smallpox and its treatment in the Republic of Mali. *Journal of Tropical Medicine and Hygiene* **28:**224–228.

Institute of Medicine. 1999. *Assessment of Future Scientific Needs for Live Variola Virus*. National Academy Press, Washington, D.C.

Miller, G. 1957. *The Adoption of Inoculation for Smallpox in England and France*. University of Pennsylvania Press, Philadelphia, Pa.

Preston, R. 1999. The demon in the freezer. *The New Yorker* 12 July 1999, pp. 44–61.

Soper, F. L. 1966. Smallpox—world changes and implications for eradication. *American Journal of Public Health* **56:**1652–1656.

Tucker , J. B. 2001. *Scourge: the Once and Future Threat of Smallpox*. Atlantic Monthly Press, New York, N.Y.

Watts, S. 1997. *Epidemics and History: Disease, Power and Imperialism*. Yale University Press, New Haven, Conn.

Polio

Daniel, T. M., and F. C. Robbins (ed.). *Polio*. University of Rochester Press, Rochester, N.Y.

Donnelly, J. 2000. A dream deferred. *Boston Sunday Globe* 24 Sept. 2000, pp. A1–ff.

Gould, T. 1995. *A Summer Plague: Polio and Its Survivors*. Yale University Press, New Haven, Conn.

Hays, J. N. 1998. *The Burdens of Disease: Epidemics and Human Response in Western History*. Rutgers University Press, New Brunswick, N.J.

Hinman, A. R., et al. 1988. Live or inactivated poliomyelitis vaccine: an analysis of benefits and risks. *American Journal of Public Health* **78:**291–303.

Le Comte, E. 1958. *The Long Road Back: the Story of My Encounter with Polio*. Victor Gollancz, London, United Kingdom.

Lockhart, J. G. (ed.). 1837. *Memoirs of the Life of Sir Walter Scott, Bart.* (2 vols.), vol. 1, p. 23–27. Carey, Lea, and Blanchard, Philadelphia, Pa.

Paul, J. R. 1971. *A History of Poliomyelitis*. Yale University Press, New Haven, Conn.

Remignanti, D. 2001. An audacious effort. *Dartmouth Medicine* **Spring:**44–54, ff.

Schrope, M. 2001. Plans to eradicate polio hit by virus outbreak in Bulgaria. *Nature* **411:**405.

Seavey, N. G., J. S. Smith, and P. Wagner. 1998. *A Paralyzing Fear: the Triumph Over Polio in America*. TV Books, New York, N.Y.

Sever, J. L., et al. 1992. "PolioPlus," a booster shot. *World Health Forum* **13:**10–14.

Smith, J. S. 1990. *Patenting the Sun: Polio and the Salk Vaccine*. Anchor Books, New York, N.Y.

World Health Organization. 2000. *Global Polio Eradication Initiative, Strategic Plan, 2001–2005*. World Heath Organization, Geneva, Switzerland.

World Health Organization. 2000. Special theme—polio eradication. *Bulletin of the World Health Organization* **78**(3)**:**281–410.

General

Andrews, J. M., et al. 1962. The philosophy of disease eradication. *American Journal of Public Health* **52:**110–118.

Cockburn, T. A. 1961. Eradication of infectious diseases. *Science* **133:**1050–1058.

Dowdle, W. R., and D. R. Hopkins (ed.). 1998. *The Eradication of Infectious Diseases: Report of the Dahlem Workshop on the Eradication of Infectious Diseases*. John Wiley & Sons, Chichester, United Kingdom.

McNeil, W. 1977. *Plagues and People*. Anchor Books, New York, N.Y.

Venediktov, D. 1998. Alma-Ata and after. *World Health Forum* **19:**79–86.

World Health Organization. 1998. Global disease elimination and eradication as public health strategies. *Bulletin of the World Health Organization* **76**(Suppl. 2).

World Health Organization. 2002. *Scaling Up the Response to Infectious Diseases: a Way Out of Poverty*. World Health Organization, Geneva, Switzerland.

Yekutiel, P. 1981. Lessons from the big eradication campaigns. *World Health Forum* **2:**465–490.

Index

Q

R